Avoiding the
Extinction of Humanity
A Practical Plan

by John M. Goodman, Ph.D.

BONUS
BENEFIT:
THE ONLY WAY
TO DO REAL,
SPACE TRAVEL
FOR ALL!

Waterfront Digital Press

2055 Oxford Ave., Cardiff, CA 92007

Avoiding the Extinction of Humanity
A Practical Plan

First Edition ISBN 978-1-939116-87-1

(Also available as an ebook, ISBN 978-1-939116-86-4)

Contents

A Matter of Perspective

Saving humanity from self-inflicted extinction will be the most valuable immediate outcome of the new global infrastructures proposed in this book. But ultimately, having a way for anyone, no matter how poor, how old, or how unfit they may be, to travel into space and see our world from the outside—this may well turn out to be what future generations will see as the most valuable outcome in the long run. I base this statement on some comments by several astronauts about what that sort of trip meant to them[1].

Earthrise Photo taken by Apollo 8 crewmember Bill Anders on December 24, 1968, showing the Earth seemingly rising above the lunar surface. Note that this phenomenon is only visible from someone in orbit around the Moon. Because of the Moon's synchronous rotation about the Earth (i.e., the same side of the Moon is always facing the Earth), no Earthrise can be observed by a stationary observer on the surface of the Moon. (Original NASA image—here slightly cropped)

- The Earth was small, light blue, and so touchingly alone, our home that must be defended like a holy relic. The Earth was absolutely round. I believe I never knew what the word round meant until I saw Earth from space. - *Aleksei Leonov, USSR*

- Looking outward to the blackness of space, sprinkled with the glory of a universe of lights, I saw majesty—but no welcome. Below was a welcoming planet. There, contained in the thin, moving, incredibly fragile shell of the biosphere is everything that is dear to you, all the human drama and comedy. That's where life is; that's were all the good stuff is. - *Loren Acton, USA*

- For the first time in my life I saw the horizon as a curved line. It was accentuated by a thin seam of dark blue light—our atmosphere. Obviously this was not the ocean of air I had been told it was so many times in my life. I was terrified by its fragile appearance. - *Ulf Merbold, Federal Republic of Germany*

- Suddenly, from behind the rim of the moon, in long, slow-motion moments of immense majesty, there emerges a sparkling blue and white jewel, a light, delicate sky-blue sphere laced with slowly swirling veils of white, rising gradually like a small pearl in a thick sea of black mystery. It takes more than a moment to fully realize this is Earth … home. - *Edgar Mitchell, USA*

- A Chinese tale tells of some men sent to harm a young girl who, upon seeing her beauty, become her protectors rather than her violators. That's how I felt seeing the Earth for the first time. I could not help but love and cherish her. - *Taylor Wang, China/USA*

---

[1] http://homepages.wmich.edu/~korista/astronauts.html

Overview

The good news I bring you is that YES, we can avoid our imminent extinction—if we take an appropriate action, and do so very soon. The even better news is that if we do that in the way I propose, everyone's life will become less expensive and far better in manifold ways.

We are at a crisis point. There are so many humans on the planet now, and we each use so much energy, with almost all of it coming from burning "stuff" (fossil fuels and other things), that our activities collectively are now driving our global climate. Largely unaware of how much this is true, we've been running amok. Within approximately the next century, a great many species that now are thriving will be extinct—and that probably includes us. Unless, that is, we choose to respond to this crisis appropriately and quickly enough.

All of the suggestions so far for how we can avoid this end are simply inadequate. Following them will slow things down somewhat, but they won't stop the coming extinction.

Think of it as driving off a cliff. If you start doing that, your passengers may suggest that you slow down, but that will only delay the disaster. Or they may say it isn't yet clear that this really is a cliff, so let's just go on until it becomes clearer. Waiting until you are already falling will only make it impossible to recover. What you have to do is turn aside. Take a new path. (Well, slowing down *while* you turn in a new direction *is* likely to be a good idea. But not *just* slowing down.)

We need a bold new plan—one that is both feasible and effective. This book sets out just such a plan.

Specifically, I propose that we build a set of global infrastructures that will let us harvest all the energy we can ever possibly use directly from the sun in a way that is so much more effective than anything we do today that it will make that energy at least ten times cheaper than the cheapest source we now have (the most efficient modern natural gas power plants). Having this new low-cost energy source will enable us to change how we meet all of our current energy needs, moving away from using energy derived from burning stuff to using only electricity directly from the sun. The low cost of this new solar energy will ensure that this conversion happens rapidly.

Transportation now uses about a third of all our energy, and some have argued that things like ships and airplanes need liquid fuels. With my plan we will also have all long-distance transportation provided by extremely high-speed superconducting magnetic levitation (maglev) trains powered by solar-generated electricity. These trains will span the oceans and will carry both people and cargo up to one hundred times faster and at a cost that is up to one hundred times smaller than any transportation means we have today. Again, once this new system is available, market forces will rapidly drive the change to using only this system for long-distance transport.

* * *

So, what is this bold new plan?

I propose that we build a network of super-tall trestles that interconnect all of the major metropolitan areas of the world. A trestle, I remind you, is a structure whose top is at an essentially constant altitude, and whose value resides solely in the value of the things it supports.

The trestles I'm suggesting will be about 100km tall (about 60 miles). Something that tall must have a very broad base for stability—probably at least 20km (12 miles) from side to side. Given that it is going to be that tall and wide, I suggest we make the top at least 5km (3 miles) wide to enable putting many different useful things up there.

To keep these structures from being oppressive presences, I suggest that they be built as lacy, open structures that block at most a few percent of the sunlight when it is coming directly down through the structures. Most of the time it will be coming in at an angle and will cast even less of a shadow on the ground below.

In fact, it will probably be possible to build them in a way that will make almost all of their extent invisible to the eye. And they can have ample open regions underneath them in between their ground supports to allow airplanes to pass under them.

I've checked with many experts and have found ample reasons to believe that building them will be possible, although it certainly will be a major engineering and manufacturing challenge.

Building them will also be very expensive. But you should think of that as an investment in our survival as a species, in addition to being the world's largest jobs program and largest economic stimulus program ever. (Historical examples, such as the U.S. Interstate Highway System and high-speed trains in France, Spain, and elsewhere, suggest that this investment will pay back its investors very generously.)

What we will put up on top of these trestles, at least at first, are only some well-understood, commonly used, and thoroughly tested technologies.

First, we'll put up there solar photovoltaic panels that are constantly turned to accurately track the sun. These panels will be essentially outside the

Earth's atmosphere, and as such, they will see far more sunlight more of the time than they could at ground level. This is the main reason why they will be approximately ten times as effective as ground-based solar panels.

By surrounding the planet with these trestles and their solar panels and interconnecting them with a superconducting electric power distribution system, we can provide this solar power 24/7 everywhere on the Earth. Furthermore, this system will let us bring that electric power to the ground at numerous distribution points near all of the users of that power, and do so with *no* loss along the way—quite unlike most of today's long-distance power distribution means.

Next, we'll put tracks for superconducting maglev trains up there, with spurs that curve down to the ground at each major metropolis that the trestle passes. Doing this will allow us to make a global transportation system that will carry containers filled with either cargo or people from point to point on the Earth very quickly and at a very low cost. This system will carry these containers very gently (with an acceleration of no more than one-tenth of gravity) from anywhere to anywhere else within less than one hour. Sending a typical cargo container (specifically the popular 2TEU, 12m [40ft] long units) on even the longest such trip would cost less than $100USD no matter how heavy the load it contains. The cost per one-way passenger trip would be less than $35USD.

A lovely bonus to this is that we can also use this system to build the first true mass-market space craft launch and retrieval system. Anyone who wishes to travel into space will be able to do so with no great cost nor any requirement to be astronaut-fit.

Finally, we'll use an aspect of the solar power-generating system—possibly expanded to make it more effective—to provide a direct means to control our global climate. This will let us compensate for the excess carbon dioxide we've already put into our atmosphere. And, in addition, this will provide a means to modify what are now global-climate-change exaggerated local weather patterns by reducing the temperature differentials that drive that extreme weather.

* * *

Just over half a century ago the United States of America faced a challenge. It was the Cold War and the USA feared it was losing the space race to the Soviet Union. It needed a bold new plan to solve that problem.

Some scientists and engineers had talked about maybe someday sending rockets with people all the way to the Moon. But they didn't know how, exactly, we could do that.

President John F. Kennedy decided to take a bold step. He announced that we would send a man to the Moon and return him safely and do so within the decade. He also said it would be hard, and cited that as one good reason for doing it. It would stretch us, challenge us to do our best. And we would succeed.[2]

It began a very exciting time in the country's history. And the initiative was a success. Two men were put on the Moon and returned to Earth safely, and that happened before the end of the decade.

That was merely a crisis of national pride. Today all of humanity faces a much more serious crisis. Truly an existential crisis. We need to take a similarly bold step, accept the difficulty and expense, and solve it as quickly as we can.

The engineering for the challenge I am proposing, plus building the industrial processes that would be needed, might take most of a decade or maybe even two. The building of the full system could take several more decades, although once we see the benefits begin to emerge I suspect we'll put on more speed and finish the job rather more quickly than we can now imagine.

How This Book is Organized

I have divided this book into six main parts. Here is a brief summary of those parts.

The first explains why humans need so much energy, how we get it, why this is driving our global climate, and why none of the proposed "solutions" to this problem is sufficient.

The second section describes the super-tall trestles I am proposing that we build. It also tell you why I think we can do that, even though I do not know, at this point, exactly how we will do it.

The third section describes the new global solar power generation and distribution system. Here you will learn why it will be so much more effective than ground-based solar power.

The fourth section describes the transportation system in enough detail that you'll be able to visualize how it will be able to carry both cargo and passengers so quickly and inexpensively.

The fifth section show how a clever use of the solar power system can also let us directly control our global climate—and even modify how sunlight

---

[2] His actual words are on page 104.

power is distributed over the globe, possibly letting us minimize the extreme weather our current global climate is producing.

The sixth section provides some additional thoughts and a summary.

Before you plunge into one of these parts, I thought you might want to know a bit more about who I am and how I came to write this book. (My more formal biography is at the end of this book.)

A Personal Note

Who, you might wonder, is making these audacious suggestions? Let me introduce myself.

I am not a climatologist (that is a scientist who has specialized in studying the science of climate history and change). I was trained as a physicist. Specifically, I got my B.A. in Physics from Swarthmore College, and my Ph.D. from Cornell University with a major in Experimental Physics, with minors in Theoretical Physics and the History of Science and Technology. My thesis research was in low-temperature solid-state physics. I've always been interested in learning for myself as much as I could about almost any branch of science or technology, a trait that has, as it happens, helped me be pretty good at what I have spent most of my life doing.

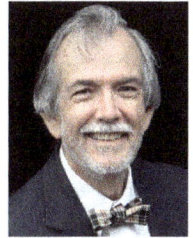

I have not, primarily, been a research scientist—working to extend scientific knowledge. Instead, I have been primarily a teacher—helping others understand what scientists and engineers have accomplished and what is still unknown, and perhaps helping them realize where we should be focusing our attention and efforts.

I've had over five decades of teaching experience in many different venues. I've been a classroom teacher at various colleges and universities, a high school, and several proprietary schools, and for several years I was a course director and taught multi-day classes on the road for a technical seminar company. Some less obvious "teaching" roles have included being a consultant to various small to medium-sized businesses, being the founder, exhibit designer, and Executive Director of an interactive science museum, and being the author of seven books explaining how personal computers work.

I never set out to study climate change or to write a book about our impending self-caused extinction. That was the accidental result of something very different. It began when I asked myself a question and then began to do research to see if I could find an answer.

My journey began with my personal desire to enjoy the experience of space travel. I know that I'll never do it the way it is now being done. I'm too unfit, and not nearly wealthy enough. So I wondered if there might be a way to do space travel differently. A way that would let *anyone* experience that sort of adventure, without having to be particularly fit or wealthy.

I came up with an answer—a way that could, I believe, be built and provide just that sort of experience at a very modest cost (to be more precise, a very modest "marginal cost"). [The **marginal cost** means the cost for each additional person's travel, as opposed the **capital cost** to create the system in the first place.] However, I realized that creating this system would be so hugely expensive that I doubted there would be enough market for its services to motivate anyone to build it.

Then I realized something: This innovation would have many other, much more valuable uses. To begin with, it would revolutionize long distance travel. Maybe that would make it more attractive to investors.

I also knew that this system would need a lot of energy, so I looked into how that might be supplied. I found a way to do that and this turned out to be even more revolutionary in its potential impact. We could get *all* of the energy that all of humanity could possibly use for any purpose from the sun nearly for free, once the system itself was built.

Further, I realized that if we could do that, moving away from fossil fuels to this alternative energy source would be a no-brainer. Instead of trying to get people to sacrifice and pay more to be "green," this would let us all feel good about how much money we were saving even as we felt good about not doing bad things to our environment.

So I began to study the impact of burning fossil fuels for energy, which is how we now get over 90% of our energy. **And what I learned scared me!**

Along the way I also learned a lot about how our world population has grown—and is still growing—and how it might be curbed, plus about some other population-related problems that some countries are now facing (and many more will soon be facing) when dealing with a declining population. I realized that the same innovation I came up with that can save us from extinction, and give us "free" energy, and more could also help relieve all of these population-related problems as well. That is the subject of a companion book I'm calling *Multiple Population-Related Problems and a Surprisingly Graceful Solution.*[3]

---

[3] *Multiple Population-Related Problems and a Surprisingly Graceful Solution*, by John M. Goodman, Ph.D. Waterside Digital Press, 2014 (ISBN 978-1-939116-88-8 as an ebook or 978-1-939116-89-5 for a print on demand paperback)

I wrote this book because I believe we are in serious trouble and what I've imagined—expensive and difficult as it will be to create—may just be the most practical solution we can find to our present problems.

A Peek at Our Possible New Future

One morning, in Los Angeles, in California, in the United States of America, a young man named Glen and his wife of only three years, Peggy, are having breakfast. Glen excitedly says to Peggy, "Honey, I took off work today. I want to do something special with you." She's interested, so he goes on, "I thought we might like to go to dinner and take in a show in Paris, France tonight."

"Can we afford that?"

"Easily. I was looking around online last night, and I figured it all out. We can travel to Paris and back for less than it costs for dinner there and the show. However, if we want to do it today, we'll have to leave soon. After all, it is already early afternoon there." [4]

"Okay, I guess. What'll I wear?"

"Better dress up. After all, we're going to go out on the town—and not just any town. We're going to Paris!"

Glen spends the next few minutes online securing reservations (with a total trip cost that really is less than the cost of their dinner and the show they'll be seeing), while Peggy gets ready. Finally, shortly after 8AM, they go to their neighborhood transportation center. There they get into a special train car with perhaps a couple of dozen other folks, all of whom are also going to Paris, France.

This train car is unlike the commuter train Glen normally takes to work. This one is set up more like an old-fashioned "parlor car." Instead of rows of seats—which Glen thinks of as pews, like in their church—this one has several conversational groupings of chairs. Some are around small tables; others have a low coffee table in the middle; yet others are just high-backed wing chairs for folks who want some privacy as they read. The car has no windows, but it does have several large view screens which will mostly show what one would see if you looked out a window in that location: on either side, at the front and rear of the car, or up at the ceiling. There is even one in the floor—currently showing the track beneath the train.

Everyone settles in.

---

[4] Paris is eight or nine hours ahead of Los Angeles, depending on whether or not Daylight Savings Time is in effect.

The train door is closed and locked. From the view on the screens the passengers can see that the train is moving. A steward asks if anyone would like a beverage, and then provides whatever people request.

After awhile, Glen notices a group of men around a card table across the car from them drinking and playing cards. They are speaking French. He asks his wife, "Honey, who do you think those men are?"

She replies, "I'm not sure. Let's ask the steward."

They beckon to the steward who comes over to them and asks, "Is there something I can do for you?"

"Yes. We're wondering who those men are over there. With their ties pulled down and their jackets off, drinking beers, they look like they are businessmen at the end of a work day—but it isn't even 9AM."

"Oh, yes. I know them well. They are regulars on this run, and it is nearly the end of their work day. They are salesmen for a huge multinational corporation. They come to Los Angeles once a month for an early morning meeting with their fellow salesmen from all around the globe. LA is a convenient place for these meetings.

When these men left Paris earlier today it was early afternoon their time. Likewise for their colleagues from London and all over Europe and Africa. Their colleagues from our East Coast as well as ones from various places in South America, including Buenos Aires, Argentina and Rio de Janeiro, Brazil left home for this meeting a little after their normal breakfast time. Their Los-Angeles based colleagues merely had to get up a bit earlier than usual. The representatives from India had to leave home after dinner (today!), while the ones from Moscow left in mid-afternoon (today). The only hardship cases came from East Asia (including China and Japan) or from Australia. They had to leave their homes in mid-evening today to make it here for a 6:30AM (Los Angeles time) meeting today.

"Now that their 'breakfast' meeting (which is what it was for the folks from anywhere in North or South America) is over these gentlemen are on their way home. They, like you, will get to Paris in time for an early evening dinner. Their North and South American colleagues are already launching into their normal day's work after their breakfast meeting. The folks from Russia and India are finally going home to bed—just a little later than their normal bedtimes. Those hardship cases from East Asia mostly choose to get a good night's sleep in Los Angeles before heading home. Otherwise they'd arrive home some time after midnight according to their local time."

"Oh," Peggy replied. "I hadn't thought of that. Thank you."

The rest of the trip is pretty uneventful in terms of what happens inside the train car. Mostly the passengers just sit and read, talk, and look at the view screens. (Some folks are watching TV on their personal tablets.) The images on the big view screens, however, become quite stunning during the middle of the trip, showing many views of the earth from high above the surface as well as views of outer space. Peggy comments to Glen at one point, "These views are really hard to believe. I feel as though I am just sitting in our living room, and yet we are apparently sailing through space, just outside of the earth's atmosphere. Incredible!"

"Well," Glen replies, "It is rather remarkable. But, you know, many thousands of people take this sort of trip every day. It only used to be impossible to do this sort of thing. Back before the GEG and HST Systems changed everything."

Something between one and two hours after they entered this car, it stops and the steward announces that they have arrived in Paris at one of the main downtown train stations.

All of this time the passengers have experienced nothing more than the sort of gentle acceleration you'd experience in an elevator in a high-rise office building—and less than the forces you experience in a automobile rapidly going around a corner. Yet, they've managed to travel just over 9,000 kilometers (about 5600 miles)! And actually, about half of the time was spent traveling rather slowly on ground-level tracks as the train car was carried from the couple's neighborhood transportation center to the Hyper-Speed Transport uplink facility for Los Angeles, and again from the Hyper-Speed Transport downlink facility for Paris to the downtown Paris train station. Only about three-quarters of an hour was spent on the Hyper-Speed Transport System, and during that time almost all of the distance on this trip was traversed, with the train's peak speed being approximately ten kilometers per second (about 22 thousand miles per hour)—or nearly fast enough for a space ship to completely escape earth's gravity.

The couple go to dinner and see a show, spending seven hours in Paris before they return to the train station. After that, they are back home before 7PM, in plenty of time to watch the evening television news, or simply go to bed after what you and I would consider a really exotic day's adventure, but which this couple, and many, many others like them in this not-distant future, see as just another moderately special outing. Something they can easily do several times each year.

And tomorrow Glen will be back at his job in Los Angeles, just as usual.

Part One
Driving Off the Climate Cliff

Some Relevant Context...and some Hope

There have been five great (and at least another five not quite so great) mass extinction events in our planet's history.[5] The most devastating one happened 250 million years ago. That one caused the extinction of 90% of the species then living on the Earth. The most recent one was 65 million years ago. That one caused the extinction of about 62% of all the species, including the dinosaurs.

As far as we can tell, each of them was caused by a global climate that was stressed, plus some "sudden" event (in geological terms) that triggered the collapse. [Sudden here means something that might happen in a day, or in a year, or at least within one or two centuries. Most "normal" geological changes take thousands of years to happen, so anything much faster than that qualifies as "sudden."] Perhaps the most famous was the huge space rock (the Chicxulub meteor) that crashed into the Earth near what is now the Yucatan peninsula 65 million years ago.

---

[5] For a general discussion of this topic look at
http://ethomas.web.wesleyan.edu/ees123/mass_extinctions.htm.

Species go extinct all the time. But in normal times the rate at which they go extinct is a fairly constant "background rate."

The observed rate of species extinction now is around 70 times faster than the background rate seen over long geologic periods in the past, and is comparable to that seen at the start of each of the great mass extinction events in the past. For this reason, many scientists are beginning to talk about a sixth great mass extinction event. And they make the distinction that, while it appears that many of the previous five were triggered by some external influence, this one can be laid squarely at *our* feet. Human activity is causing the rise in species extinction now, and may soon tip it over into another great mass extinction event.

This potential disaster may well be coming far sooner than you might think. In fact, many people now alive may well be among the last human beings on the planet.

A Note of Hope

I wouldn't be writing this book if I didn't have hope. I am very much concerned about the impending danger. And I doubt any of the other proposed "solutions" will actually solve anything in the long run.

But I have hope because I have, quite accidentally, stumbled on what may well be our last best option for really solving this problem. Indeed, that is what this book is mostly about: Explaining that Real and Sustainable Solution to our present quandary.

It's a Dangerous Road We've Been Traveling

We've been going down this road for quite awhile now. The warning signs are obvious, at least to those who know what to look for and believe what they see. Today's signs aren't the only ones. We've seen them before, but have ignored them. The road looks so nice. Surely the signs must be mistaken.

What exactly does this sort of sign mean? It means some authority has looked ahead and has seen a danger so great that travelers should be kept from traveling past here. That raises a whole lot of good questions:

Some Serious Questions

Who is this Authority? What is the danger they have foreseen? Do I believe them? Do they have the "authority" to keep me from continuing on this road, if I want to do so? If I do go on until I see for myself what the danger is, can I then retrace my steps until I get back to where I can turn off for another road that may lead me around the danger? Is there any such alternate road?

Some Answers to Those Serious Questions

In this case, the principal authority is a huge number of climate scientists who, under the banner of the Intergovernmental Panel on Climate Change, have now issued five "Assessment Reports" over the past twenty-some years.[6] In each of them they've said we are heading for disaster—with each subsequent report refining the science behind this claim. As a cautious group of scientists they have sought to publish only the most solidly supported science, vetted by a consensus process with hundreds or thousands of highly-qualified scientists having signed off on each point.

And, no (regrettably), they don't have the authority to stop people from traveling on this road. All they can do is post the signs and hope people read them and respond to them.

The IPCC's conclusion couldn't be much clearer: Human activity has produced substantial change in our global climate. Change which is ongoing and will in the very near future lead to climatic conditions never before encountered by humans at any time in our roughly 200,000 year history as a species.

Many people are uncertain as to why we should believe this. They don't see the steps by which we have risen from a minor species on the planet to the un-challenged top predator, the dominant species, and now the one that is about to destroy what we have found to be a very comfortable environment.

Some people ignore the signs because they just don't believe in science. They aren't swayed by what they are told are the facts. They rely on faith or on their own perceptions, which, so far, haven't confirmed what the

[6] You can find all of their reports, plus information on how the IPCC reviews the data at http://www.ipcc.ch/index.htm.

scientists are telling them. The cartoon I've included here shows how one editorial cartoonist[7] put it.

Besides, they say, what else can we do? We've got to keep moving forward, and it isn't clear what other road we might take.

Other people are all too aware that there is danger ahead, but they aren't willing to pay now what they are told is a necessary price to do what will save us from that danger in the future. Their focus, like the other group I just mentioned, is short term…focusing on what they can see in front of them. On they are focused on making it through yet another day, another quarter, or another year.

Even most of the people who are aware of the danger, believe that continuing our present behavior will merely make life a bit more uncomfortable in the near future (because of human-induced global climate change). They think we can just move ourselves and our farms to higher latitudes and/or altitudes to make up for higher average Earth temperatures.

They don't realize the very real likelihood that we are fast approaching some irreversible tipping point which could then cause an inescapable mass extinction event. In other words, if we wait until everyone is clear about the problem, we may not then be able to "retrace our steps" and take some alternate path around the danger. Once you drive off a cliff, it generally isn't feasible to climb back up and turn to go in another direction.

How Did We Create This Problem?

To understand the answer to this problem it helps to divide it into three distinct parts or aspects. First is our hunger—or even our need—for a lot of energy per person, second is the sheer numbers of people, and third is the way we have historically gotten that energy, and which is still the way we get it today.

In this book I shall explain in some detail the first and third of these aspects, and just summarize the second one. I have written another, companion volume that goes into the second aspect in depth, and explains why the new infrastructures I am proposing will help us solve various problems associated with population growth or decline as well. (See the footnote on page 6.)

Aspect One

All animals need energy to do what they do. Just living takes energy, to metabolize our food, to think, keep our bodies working, and to take voluntary

---

[7] David Horsey, Los Angeles Times, February 13, 2014.

action. The bigger the animal, the more energy it needs just to nourish all the cells in its body.

Most animals are only able to use the energy they get from metabolizing the food they eat. Humans are very different.

A typical human eats between 2,000 and 4,000 nutritional calories per day. That works out to be an annual amount of something less than two megawatt-hours (2MWhr), using the equivalent unit for electrical energy. But the average human on the planet today uses roughly ten times that much energy. Even the poorest among us uses that much food energy plus about that much additional energy in some different form.

Energy From the Food We Eat

4,000 nutritional calories per day
= 4,000 kilocalories per day
= 4.652 kilowatt-hours per day

X 365.25 days/year =
1.70 Megawatt-hours per year

People living in the developed countries of the world use even more energy than that. On average they use about 33MWhr of energy per person per year. The most energy-dependent countries (most prominently the United States and Canada) use at least 80MWhr of energy per person per year.[8]

That has been our first step toward disaster. We use a whole lot of energy. But using all that energy has also been a *necessary* step toward our thriving and achieving our full potential.

Aspect Two

The second aspect of our problem is how many of us there are. Humans have taken over nearly every possible ecological niche on the planet. At least every place on dry land. And we've been a fertile bunch. There are now over 7 billion of us, and demographers predict that by 2100 there will be at least 11 billion of us.

Putting that in energy-usage terms, we now use as much energy as roughly 70 billion human-size animals (Multiply 7 billion people by the average human's use of about 20MWh per year, then compare that to an average non-human animal of our size that uses only about 2MWh per year from the food it eats). And by the end of this century that will likely grow to as much as somewhere between 180 and 450 billion human-size animals.

---

[8] Data from International Energy Agency at <u>http://www.iea.org/stats/index.asp</u>.

Fortunately, the rate at which the world population is growing has begun to decline. If, and only if, all of the developing nations of the world finally achieve full development, we can expect that rate of growth to fall to zero...or possibly even start a gradual decline in the total world population.

Perhaps you've already read about this topic and are concerned, not so much with the possibility that our world population will rise too high, but rather with the very real possibility that once we learn to limit our growth, we may do that so "well" that our world population starts to decline, perhaps even rather sharply. If so, you may be worried about too few workers to support too many retirees (which is a problem in several countries right now) or perhaps you believe that we need to have population growth to experience a healthy and growing economy. To both concerns I say, don't worry. There is a way we can have a vital and growing economy and a thoroughly acceptable safety net for older people even with a declining population that has far fewer workers than retirees. That is the focus of my companion book, *Multiple Population-Related Problems and a Surprisingly Graceful Solution* (see footnote on page 6).

Aspect Three

The final aspect of our problem, and the final step toward our disaster is how we have gotten all that energy.

Over time we have used a number of means to provide ourselves with the energy we needed in order to thrive to our full potential. At first we co-opted the energy of others. Think using oxen, horses, and other animals, also children, and—for the powerful few—the use of slaves or servants.

Then we added energy from other, non-animal sources: Fire (mostly burning branches of downed trees), water power, wind power, and later on burning peat, coal, oil, and natural gas. More recently we've added geo-thermal power, and solar power (both solar-thermal and solar-photovoltaic). But mostly we still get the bulk of our energy by burning stuff—organic stuff (carbon-containing materials).

Burning anything organic generates greenhouse gases. Even our metabolism combines organic matter with oxygen from the air and creates as byproducts methane (think of "breaking wind") and carbon dioxide plus water vapor. Burning stuff outside our body creates carbon dioxide and water vapor. Because of the much higher temperatures involved in this sort of external burning—and the fact that oxygen is only one-fifth of air, with nitrogen making up nearly all the rest—this sort of burning also produces oxides of nitrogen.

Methane, carbon dioxide, the oxides of nitrogen, and water vapor are all powerful greenhouse gases (GHGs). The more of them there are in our atmosphere, the warmer the Earth will become.

In case you didn't fully understand what I just said, I'm going to review briefly some basics about climate science. (If you know all about the physics and chemistry of climate science and about the history of our planet's climate, feel free to jump to the section titled "Where We've Been Getting our "Extra" Energy" on page 30.)

The Basic Physics of Climate

In space there is only one way that energy gets from one place to another: radiation. Electromagnetic radiation, to be precise. That is just another name for pure energy that is on the move. When it happens to be at the frequencies our eyes respond to, we call it light. When it has a much longer wavelength (lower frequency) we call it radio waves. At much higher frequencies we call it X-rays.

Our eyes only respond to about an octave of light frequencies. That is, the frequency of blue light is about twice that of red light. Light that is "bluer than blue" we call ultra-violet. Light that is "redder than red" we call infrared.

Any object that is hot radiates energy. The hotter the object the more energy it radiates, and while that energy comes off at all frequencies, there is a dominant frequency which becomes higher the hotter the object.

The sun's dominant frequency is right in the middle of our visual range. (This is no surprise. Evolution equipped us with eyes that make optimal use of the sun's light.) When you see a hot glowing coal in a fireplace, that coal is approaching the temperature of the sun, but it is still significantly cooler, hence its light is noticeably redder. Another object that is hot, but not quite that hot, will radiate mostly in the infrared, where our eyes cannot see it.

The sun shines on the Earth. Some of that light bounces off. The rest is absorbed. When the Earth absorbs energy it gets hotter. And so it radiates energy away. Once the amount of energy that leaves the Earth each second equals the amount that arrives, the Earth's temperature stabilizes. Thereafter, if either the amount of energy arriving changes or the amount leaving changes, the temperature of the Earth will change until those energy flows are again in balance.

That's the most basic statement of the physics behind climate change. But there is more to the story that you need to know.

Our Earth is surrounded by an atmosphere. Some of the sunlight bounces off the outside of that atmosphere. About half of it gets at least partway in. Somewhat less than half of it makes the trip all the way down to the ground where it is either absorbed (by the darker regions) or reflected (most especially by snow or ice). The rest of the sunlight started into the atmosphere but got absorbed there before it could get to ground level.

The bulk of the radiant energy coming up off of the Earth is at a much lower frequency (longer wavelength) than the bulk of the sunlight coming in, just because the Earth is so much cooler than the sun.

Our atmosphere is about 80% nitrogen, 20% oxygen and just trace amounts of everything else. At present about 400 parts per million of it is carbon dioxide. That is to say 0.04% by volume is carbon dioxide. Surprisingly, that number is pretty constant all over the globe, with the exception that it is higher right near fresh sources of carbon dioxide.

The amount of water vapor in the air varies by location and time a lot more than the amount of CO_2 does. There are also other important trace substances in our air that I won't go into here.

Now the killer additional fact is that our atmosphere is transparent to visible light (which is where most of the energy from the sun is). Pure air (just oxygen and nitrogen) is about equally transparent to infrared light, but when it is contaminated with greenhouse gases (GHGs), it is not at all transparent to light in the infrared, yet still quite transparent to visible light.

For any particular level of GHGs in the atmosphere (and carbon dioxide is the most important one, at least over the long term) there is some global average surface temperature for the Earth that will cause an amount of energy to escape from the planet that just exactly equals the amount that arrives at the planet's surface from the sun.

Change the level of GHGs, and the temperature must change over time until things are back in balance. When you turn up the flame on a stove under a pot of water, the temperature of the water doesn't immediately jump up. It takes some time before it responds. In the same way, if the GHG level rises suddenly, the planet's temperature won't rise immediately—but it will eventually.

Also important to know is that another factor comes into play. If we pollute the atmosphere with anything that makes it reflect more visible sunlight, that action can lower the amount of sunlight energy that gets down to the ground. That can

cool the planet. (But it also can have some other, rather less desirable consequences, like making the air stink and our eyes sting, and perhaps it might make it impossible to see the stars at night or ever have a blue sky overhead in the daytime.)

Finally, if we alter the reflectivity of the ground, that too can modify the balance. The most important example of that for our purposes is that if the Earth warms up enough, then all the ice in the sea and the glaciers on land will melt, and the Earth will, overall, look darker. That will make it absorb more sunlight and that, in turn, will make it get even warmer.

This last point is very important. It is a positive feedback that can make what is otherwise just a small change into a "tipping point" phenomenon where even a little bit of temperature increase could cause a much larger temperature increase later on. Climate scientists refer to my first example (polluting the atmosphere) as one of the fast-feedback response mechanisms. They call things like the melting of the glaciers part of the slow-feedback responses to a warming of the planet.

The Basic Chemistry of Climate

There are three really important chemical reactions you need to know about. One is photosynthesis, another is simple combustion, and the third is a variant form of combustion. The first one takes energy from the sun and, in effect, stores it. The other two release that energy once again.

Photosynthesis is the process by which plants (and before them some very old bacteria) create organic matter, including the leaves and trunks of trees, for example, or more baby bacteria for another example, out of carbon dioxide and water (either as vapor from the air, or as a liquid drawn up from the ground by a tree's roots, for example).

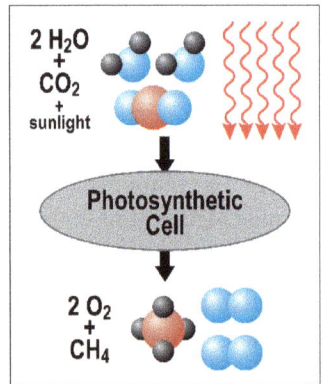

The plant (or bacterium) combines the water (H_2O) and carbon dioxide (CO_2) and, using the energy it gets from sunlight, causes the atoms of these molecules to swap places, producing new chemicals, including oxygen molecules (O_2) and methane (CH_4). The oxygen is released into the air; typically the methane is used in some additional processes to build up new leaves or create "baby" bacteria. You can see that there are four hydrogen atoms (the black balls), four oxygen atoms (the blue balls), and one carbon atom (the orange ball) going into the process, and the same number of each kind of atom coming out of it, with the solar energy (shown as wavy red arrows) now stored in the methane and oxygen molecules.

Both the oxygen we breathe, the nitrogen that make up the bulk of our atmosphere, and hydrogen are all atoms that like to hang out as couples. So, if they aren't in some other chemical compound with other kinds of atoms, they naturally form two-atom compounds with themselves, as O_2, N_2 and H_2.

Methane is the principal component of natural gas, the fossil fuel that is now the preferred material to burn in order to get energy. Methane is the simplest hydrocarbon. Put several of them together with all of the carbon atoms linked to one another and the hydrogen atoms each linked to one of the carbon atoms and you get what is called a long-chain hydrocarbon. These are some of the basic building blocks of all organic matter. Add a few mineral atoms like phosphorus, sulphur, nitrogen, or oxygen, and perhaps rearrange the carbons in rings or some other form and you can get every organic chemical known to man or nature.

Photosynthesis is the principal method by which sunlight energy is stored in plants. If those plants get crushed under enough rock and other things, and "cook" for long enough, they may turn into coal, oil, or natural gas. That's how the fossil fuels we burn today were created many millions of years ago.

Combustion is the reverse of photosynthesis. If you put together a couple of oxygen molecules and a methane molecule, add just a little energy (perhaps as heat) the result will be the creation of a molecule of carbon dioxide plus two water molecules and the release of a lot more heat.

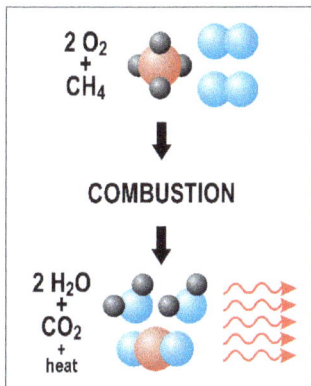

In the figure the small black balls are hydrogen atoms, the blue balls are oxygen atoms, and the big orange ball is carbon. Once again, you can see that there are four hydrogen atoms, four oxygen atoms, and one carbon atom going into the process, and the same number of each kind of atom coming out of it, plus a bunch of previously stored solar energy (wavy red arrows) being released.

Often combustion is done as a fire with an open flame. That flame can then be used to heat up something else, *e.g.* water in a steam engine, in order to put that heat energy to work somehow. Or it can simply be used to heat some air to keep a house or a person warm.

Sometimes, if the oxygen going into the combustion comes from the air (which is only 20% oxygen and about 80% nitrogen), the heat released is enough to induce some of the nearby nitrogen molecules to undergo a parallel chemical reaction in which the nitrogen atoms release their grip on one another and each of them bonds with one to three oxygen atoms. These oxides of nitrogen are also potent GHGs as well as being some of the nasty parts of smog. Another component of smog that can be produced this way is an odd form of oxygen molecule having three atoms. That form is called ozone, and it is primarily a problem because the "extra" oxygen atom in each molecule is only loosely bound, and can therefore easily jump over and attach to something else in the process "oxidizing" (burning) that other something.

If some unburned methane escapes into the atmosphere (it is a much lighter-than-air gas at normal temperatures), it acts as a super-potent greenhouse gas, but only for a relatively short time (perhaps as much as several years). Eventually, it will spontaneously "combust" with any available oxygen molecules to produce CO_2 and water vapor. Since, molecule for molecule, methane absorbs more than 20 times as much infrared radiation as CO_2, it would be the most important GHG in the atmosphere if only it didn't fairly soon turn into CO_2 and water. About half of all the CO_2 that is put into the atmosphere lingers in the atmosphere for centuries, with the rest getting dissolved in the ocean or combining with freshly exposed rocks (weathering them).

The final critical chemical reaction for you to understand is simply a low-temperature version of combustion. This reaction is extremely important for animals—like us, for example. We use it every day to extract the sun's energy that has been stored in the plant or animal matter we eat as our food. (Energy is stored in animal matter by a very similar process in which it is extracted from the plants, or other animals, which that animal ate.) This is just the first step of a process that we call **Metabolism**.

The essence of this reaction is accomplishing what combustion does without letting all of the stored solar energy escape as heat. Instead the reaction goes through several intermediate stages and in that way transfers most of that energy temporarily into some other chemicals. In the other steps of metabolism, we carry that energy—now stored in those other chemicals—through our blood to every cell in our bodies where it is then transferred into those cells to enable them to do their jobs. All of these steps proceed with minimal production of heat (and, therefore, no production of oxides of nitrogen or ozone).

Every cell in our bodies needs energy. Mostly the metabolism happens in our guts with the oxygen being brought through the blood from our lungs and the output CO_2 being carried by the blood back to our lungs to be breathed out. The energy is carried by glucose or ATP molecules to wherever in the body it is needed.

Because most of that energy never becomes heat until it is used to power our muscles or our brain cells (or the cells in our guts that perform the metabolism in the first place), it keeps our bodies from charring as they would if all that energy were released as heat where it was first liberated from the food stuffs.

The Basic History of the Earth's Climate

This is a fascinating story. And it is amazing to me just how much of it scientists have been able to learn.

The Earth and the Sun formed about the same time, roughly 4.5 billion years ago. As you might imagine, the Earth didn't look then at all like it looks now. Indeed, at first it was very, very hot. This heat came mostly from the fact that all the bits that made up the Earth had fallen together and smashed into one another to make up that whole. In the process of all that banging on one another they got so hot they melted. So the early Earth looked roughly like this picture. Just a big hot glowing ball of molten rock surrounded by a wispy cloud of vapor.

During the next half-billion years two things happened. First, the Earth cooled off enough to form a solid crust of rock, and some of the vapor condensed out as water to make up our oceans. Some of the lightest stuff in that vapor (like hydrogen and helium) escaped, since the Earth's gravitational field was much weaker than that of the Sun (which is mostly hydrogen and helium even today). The other big thing that happened is that lots of other space rocks crashed into the Earth. We call this the period of Heavy Bombardment. One particularly large space rock apparently hit so hard it splashed out a big drop of molten rock which then cooled off and became our Moon.

The Earth kept cooling down and after about another half-billion years (then 3.5 billion years before the present) one of the earliest forms of life discovered photosynthesis.

At this point the atmosphere around the Earth was mostly nitrogen with just a little bit of many other things in it, including a rather large amount of carbon dioxide. But there was almost no oxygen there.

These early life forms bloomed in the presence of plenty of CO_2, water, and sunlight to let them grow and reproduce. The first oxygen they produced got absorbed in the oceans and rocks underneath the oceans. That took about 600 million years. Then, for the next 600 million years, or so, more oxygen was created, but it got bound up in rocks on land or went into creating the ozone layer that protects us from the intense ultraviolet light from the sun.

Right after that something startling happened. These little critters had been so active soaking up CO_2 and pumping out O_2 that it got to the point that there were nearly no greenhouse gases left in the atmosphere. As a result the temperature of the Earth fell really far—it started a sort-of super ice age. At this point, about 2,300 million years ago, the Earth looked roughly like this figure, which we term the "Snowball Earth." Ice and snow covered the entire planet from pole to pole.

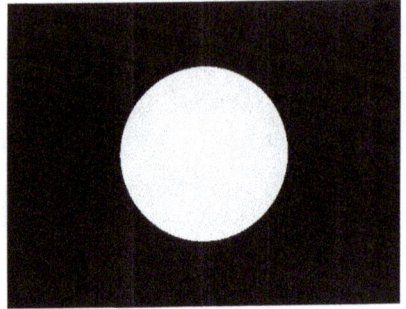

Finally, starting perhaps 850 million years ago, the amount of oxygen in the air started to accumulate, building up the concentration slowly over the next roughly five hundred million years.

And as that happened, animals started breathing it, causing some of the oxygen to be turned back into CO_2. Ever since then there has been a balance between the plants changing CO_2 into oxygen and animals changing it back. (Also, rotting plant matter emits CO_2 since rotting is yet another form of combustion.) Occasionally the balance got off and we had another couple of times, between six and eight hundred million years ago, when, once again, there was another Snowball Earth.

Eventually things settled down and life in all its forms flourished. Finally the atmosphere of the earth was stabilized at about 20% oxygen as it is now, and the earth began to look pretty much like it does now. Well, the balance was always a bit tentative and life is always threatened by something. New species kept appearing, to take advantage of some at-that-point unoccupied ecological niche. And other species kept dying (releasing their niches for yet other new species to occupy).

Every once in awhile something dramatic happened, and there was a mass extinction event that wiped out a substantial fraction of all the life forms on the planet, after which life boomed again, albeit in different forms than before.

It appears that at least some of the largest of these mass extinction events were caused by some outside influence, like a large asteroid hitting the Earth, or by some major volcanic eruption which could throw enough "junk" into the atmosphere to temporarily block the sunlight (plus perhaps dumping minerals in the ocean and changing its chemistry enough to wipe out many life forms, and thereby also killing all the life forms that depended on the now-missing life forms). Others were caused by some one of Earth's many organisms having run amok in a way that unbalanced the atmospheric composition. The worst one of all, 250 million years ago, is now thought to have been caused by just that sort of event—leading to a sudden jump in the level of carbon dioxide and a corresponding large rise in the global average surface temperature.

Sediments on the deep ocean floors record much of this history in layers as each generation's corpses fell to the floor and, over time, became fossils. This is one of the sources for our information about all of the things I've just described.

The next figure shows the history of both the global average surface temperature and the atmospheric carbon dioxide level for the past 65 million years. I've presented these data in a special manner, quite unlike what you'll find in most other sources.

Specifically, I wanted to show you that particular super-long range of history in a way that resembles how I believe most people think about time. When you think about things that happened in the past, your perspective changes with how long ago those things were. For things that happened last week, you remember (and often care about) which day they happened. If they happened a decade ago, you probably only care about which year or maybe which month they happened. If you contemplate something that happened a few thousand years ago, you aren't likely to care very much precisely which year it was. And so on.

Thus, I've drawn the time scale as proportional to the logarithm of the amount of time before some near future time. I chose 2015 as that near future time for this particular chart—although the graph only shows data for times from 2013 to the way past, because we don't yet know what the data will say for the next two years.

On the left are labels showing the time in Years Ago (ya) [that is, years before 2015], or Thousands of Years Ago (kya), or Millions of Years Ago (Mya). On the right, for the historical times, I have shown the year in either AD or BCE form. Further to the right I also have shown where a few key events in geological or human history fall on this chart.

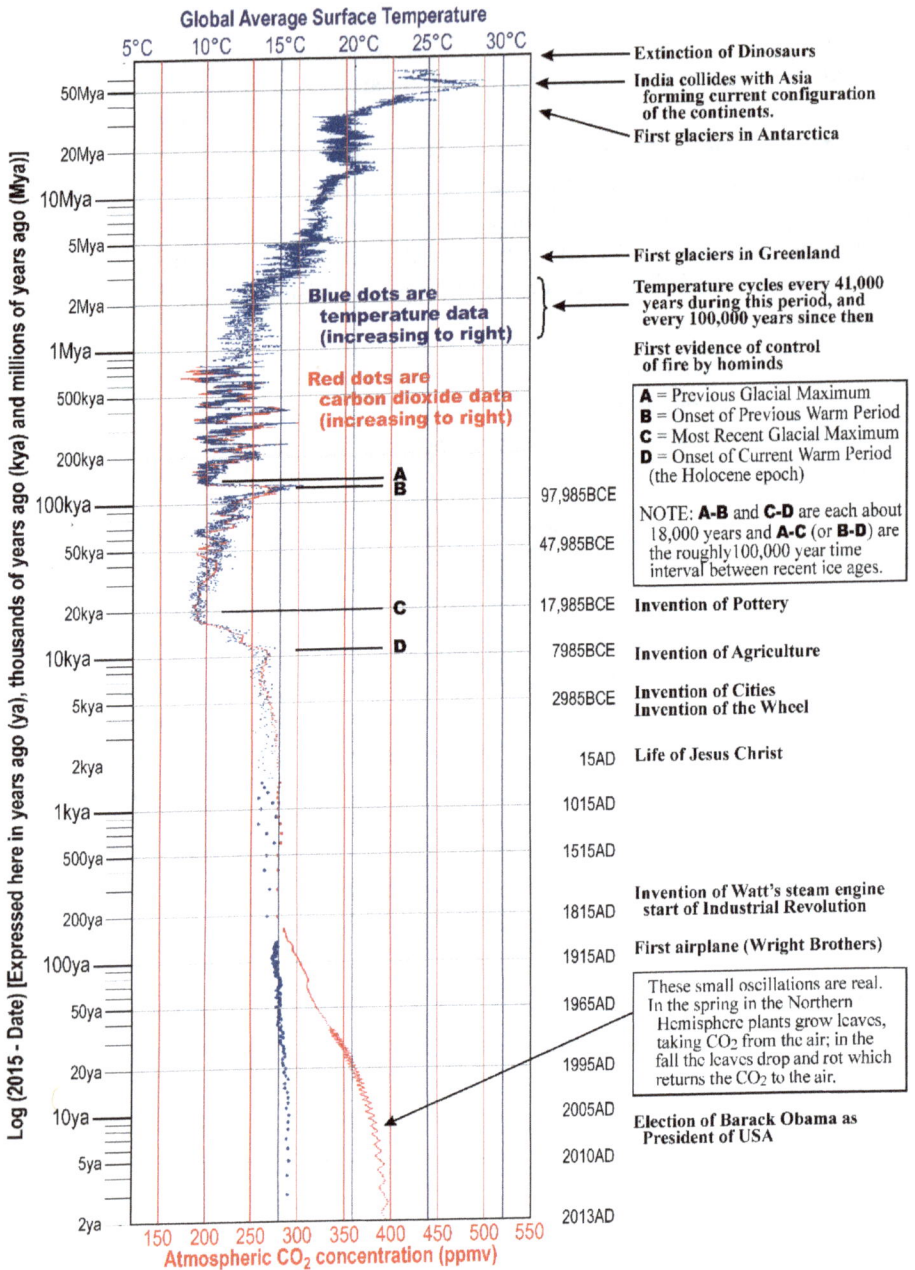

Global Average Surface Temperature

Extinction of Dinosaurs

India collides with Asia forming current configuration of the continents.

First glaciers in Antarctica

First glaciers in Greenland

Temperature cycles every 41,000 years during this period, and every 100,000 years since then

First evidence of control of fire by hominds

Blue dots are temperature data (increasing to right)

Red dots are carbon dioxide data (increasing to right)

A = Previous Glacial Maximum
B = Onset of Previous Warm Period
C = Most Recent Glacial Maximum
D = Onset of Current Warm Period (the Holocene epoch)

NOTE: **A-B** and **C-D** are each about 18,000 years and **A-C** (or **B-D**) are the roughly 100,000 year time interval between recent ice ages.

97,985BCE

47,985BCE

17,985BCE — Invention of Pottery

7985BCE — Invention of Agriculture

2985BCE — Invention of Cities / Invention of the Wheel

15AD — Life of Jesus Christ

1015AD

1515AD

1815AD — Invention of Watt's steam engine start of Industrial Revolution

1915AD — First airplane (Wright Brothers)

These small oscillations are real. In the spring in the Northern Hemisphere plants grow leaves, taking CO_2 from the air; in the fall the leaves drop and rot which returns the CO_2 to the air.

1965AD

1995AD

2005AD

2010AD — Election of Barack Obama as President of USA

2013AD

Atmospheric CO_2 concentration (ppmv)

Another unusual feature of this chart is that I have not drawn any trend lines for you. Instead I have shown you the actual data points (or in some cases the averages of several data points representing data for essentially the same time).

- 24 -

Where you think you are seeing a line you are instead simply seeing enough data point dots that they run together and appear to be a line.[9]

The blue dots are inferred global average surface temperature values, inferred from deep ocean sediments in the time period from 65Mya to about 1kya, or from ice core data from about 800kya to about two hundred years ago, or from actual temperature measurements taken at many locations around the globe and then averaged for the recent historical period.

There are no red dots for times before about 800kya. We don't have any measurements of carbon dioxide levels in air that was trapped in something that has survived from that long ago.

The red dots for the period from about 800kya to about two hundred years ago represent the measured concentration of carbon dioxide in the atmosphere found inside air bubbles in the ice cores drilled out of glaciers in Antarctica.

The red dots from 1880AD through 1979AD are taken from estimates by Dr. James Hansen and his colleagues at the Columbia Earth Institute. The data from 1979 to a bit after the middle of 2012 are the monthly averages of actual

---

[9] The global average surface temperature (Ts) and atmospheric CO2 data are taken from the following sources:

Ts from 65Mya to 1600ya from Hansen, J., et al. (2013) "Climate Sensitivity, Sea Level, and Atmospheric CO2" Phil. Trans. R. Soc. A 371: 20120294 (31 pp) with the data downloadable from this URL:
http://www.columbia.edu/~mhs119/Sensitivity+SL+CO2/Table.txt

Ts and CO2 data from 800kya to 200ya from this URL:
http://www1.ncdc.noaa.gov/pub/data/paleo/climate_forcing/orbital_variations/berger_insolation/

Ts from 1880 to 2012 from Hansen, J., et al. (2010) "Global surface temperature change" Rev. Geophys., 48, RG4004, data available at
http://data.giss.nasa.gov/gistemp/graphs_v3/Fig.A2.txt
(this gives the temperature as an anomaly from average during 1951-1980; Hansen, et. al. used 14°C as the assumed base average Ts, I found that using 15°C seemed to give a better fit with the surrounding data)

CO2 data from 1850 to 1979 Hansen, J., and M. Sato (2004) "Greenhouse gas growth rates" Proc. Natl. Acad. Sci., 101, 16109-16114.
http://www.columbia.edu/~mhs119/GHG_Forcing/CO2.1850-2012.txt

and, finally, I used the Keeling curve for 1987.9 to 2012.7, downloaded from this URL: http://scrippsco2.ucsd.edu

daily measurements at the Mauna Loa observatory in the Hawaiian Islands. (This last set of data points is also known as The Keeling Curve, after Dr. Keeling who started the program in 1956 to make these measurements at this and several other locations widely scattered around the Earth and who continued them either personally or had them taken and analyzed under his supervision up until his death in 2005.) The blue dots came from both the ice core data and data analyses by Dr. Hansen's Earth Institute group.

There are several key points that I want you to take from this figure. First, notice how closely the blue and red data align with one another. In particular, look at the time interval from about 10,000 years ago to 800,000 years ago. Here you see eight warm periods between eight "ice ages." I have scaled the two sets of data so they would align on average, but the very close way the two sets of data follow one another is a real representation of how much the CO_2 level has influenced the global average surface temperature over time.

The only place the two diverge very noticeably on this chart is in the last 200 years. During that time the CO_2 level has risen very dramatically while the global surface temperature has only risen quite modestly—around 1°C. It appears that the junk we've thrown into the air (mostly aerosols and, perhaps, smog and other kinds of air pollution), plus a bunch of sulphur dioxide (the precursor of acid rain) emitted by a number of small and unusually active volcanoes, in that time have approximately compensated for the increased carbon dioxide levels. All of this "junk" in the air increased the reflection of sunlight off of the earth—or prevented it reaching the ground, while the increased CO_2 level inhibited the radiation of heat from the ground.

But that balance is tenuous and when this sort of junk gets thrown up into the atmosphere it tends to fall back out in a matter of only a few years, while the CO_2 stays up there for many centuries.

So if we decided to try and keep this balance by adding more junk as necessary, and if we still keep pumping ever more CO_2 up there, we'll have to add more junk every year at a higher and higher cost—including the health and happiness cost we'd experience when we start getting lots more upper-respiratory health problems, stinging eyes, and gray skies—which means there also would be no more seeing the moon or any stars ever again!

Furthermore, not only are we continuing to pour lots more CO_2 into the atmosphere each year, what we've seen so far is mostly the fast-feedback responses of the global average surface temperature to the present CO_2 level. If we keep this up much longer, the slower feedback responses will kick in with a vengeance. The aerosols we've put into the atmosphere will gradually fall out of it. So, if we tried to compensate for both effects we'd have to pour a very rapidly increasing amount of aerosols into the upper atmosphere from now on, which would probably cost a whole lot—as well as being rather unpleasant at best.

Another fact you can see from this figure, although it is not so obvious in this presentation, is how the temperature of the Earth has been oscillating with one or another nearly constant length cycles for the past several million years. During this time there have been a succession of "ice ages" and in between them extended warm spells. Around the time that the first glaciers formed on Greenland (roughly 4Mya), these cycles became quite pronounced. At first the ice ages came roughly every 41,000 years. Later on (around 1Mya) that pattern shifted with the successive ice ages coming roughly every 100,000 years. Also, the depth of the ice ages (difference between maximum and minimum temperatures) increased somewhat as their intervals became longer.

An astronomer named Milankovitch has calculated that subtle effects in the Earth's orbit around the sun[10] cause the amount of incoming sunlight to fluctuate by very small amounts (around 0.05%) with two components to that fluctuation, one with a period of 41,000 years and the other with a period of 100,000 years. Presumably this small fluctuation in sunlight, amplified by the feedback mechanisms inherent in our global climate, have caused these ice age cycles during the past few million years.

According to this history we would have been about to descend into yet another ice age some time in the next few thousand years, but because of our massive burning of stuff, we are now forcing the climate to move in the opposite direction.

Look back at that figure. Notice that if you draw a line straight up from where we are now in terms of the CO_2 level, you'll not find another time where you intersect the data until you go back at least 40 million years. (There are no CO_2 data points near that time on this chart, but the close matching of the red and blue data points elsewhere suggests that they probably would have matched comparably well back where I only had temperature data available to plot.)

Forty million years ago is about when the first glaciers appeared anywhere on Earth in Antarctica. (That is, that was when the first glaciers appeared since the end of the last Snowball Earth roughly 600Mya.) It is well worth

---

[10] He took into account not only the ellipticity of the Earth's orbit and the tilt of its spin axis, but also the precession of the spin axis over time, and the resulting change in what portion of the Earth is presented to the sun at the peak of summer each year. This matters because the northern hemisphere contains most of the Earth's land mass, which is where most of the sunlight gets absorbed, and Antarctica's glaciers in the south reflect the most sunlight.

noting that just recently scientists announced that it is now clear that at least one large group of glaciers in Antarctica are already doomed to melt and/or slip into the ocean, probably in the next few hundred years.

This comparison of now and then is only looking at the CO_2 level today. Since we're still pumping out CO_2 at an ever-increasing rate, its level is going to end up quite a lot higher before the end of this century. By that time we almost certainly will have melted most or all of the glaciers all over the world, including Antarctica, in which case even more sunlight will be absorbed by the dark sea and land and we'll be cooking ourselves even faster.

Some Other Aspects of Climate Science

I have not told you all of the key things that climate scientists look at. That would take another (and rather larger) book. One key aspect I have so far not mentioned will not be easy for any of us to ignore when substantial changes in it happens. If the planet warms enough to melt a substantial amount of the ice in the glaciers on Antarctica and elsewhere, the level of all the seas on Earth will rise, perhaps by as much as 60 meters (about 200 feet). That would flood most of the world's big cities located along the coasts of the continents and change the possession of any ocean-front property from a great asset to a huge liability.

I also haven't gone into any detail about how much the oceans are becoming acidified by the CO_2 we've added to the atmosphere. They are already acid enough that coral reefs are dying, and some shellfish are having serious trouble building their shells (which dissolve in acid). Once this sort of thing proceeds too far, we may find it hard to find food, as more and more species die off. This is just one of the ways a mass extinction event can come about.

Another factor of particular importance to mammals is that we can only live in a fairly narrow range of ambient temperatures. Our bodies work hard to keep our core temperature very close to a constant 37°C (98.6°F). We constantly produce some heat from our metabolism. To keep from overheating we must be surrounded by air that is significantly cooler than 37°C to get rid of that heat, but not so cold that we lose so much energy that our core temperature falls.

Just a very few degrees rise or fall in our core temperature is fatal. If you are in a cold, but not *too* cold, environment you can put on clothes to keep in your body heat. But if you are in a hot environment, about the only thing you can do, is turn on the air conditioning—if you have some available! If the humidity is low enough, you can use evaporative cooling (sweating) as a form of air conditioning right next to your body, but that pulls a lot of water out of you as sweat, and in many places and times it doesn't work well enough.

Right now the global average surface temperature is around 16°C or 17°C. You might think that means we could tolerate quite a lot of additional global warming. If so, you'd be wrong. [11]

The average temperature in the polar regions is around 0°C. And there is substantial variation from day to night—and remember that "day" or "night" lasts six months north of the Arctic circle and south of the Antarctic circle. In the tropics, on the other hand (where the vast majority of the people on this planet live), the temperature is often in the low to mid 30s Centigrade in the daytime. Near the oceans the humidity is high, so evaporative cooling isn't useful, and the day/night temperature variation is modest. In some of the drier regions, the peak heat of the day is as high as 40°C and occasionally over 50°C. In those regions a human who goes out in the daytime without some sort of protection against the sun will very likely die of heat stroke in short order. If it were any warmer there these regions would simply be uninhabitable by humans.

As our climate warms, more and more of the land area will get too warm for humans to live there, unprotected by air conditioning, or some other means of keeping cool. That may be another way we might be driven to extinction—if

---

[11] In 2003 there was a heat wave across Europe that killed many thousands of people. The largest toll was in France, well inside the temperate zone, indeed more than halfway from the equator to the north pole. In an article at the time in USA Today which you can see at this URL: http://usatoday30.usatoday.com/weather/news/2003-09-25-france-heat_x.htm the comment was made that this sort of extreme weather rarely happened in the past, but with global warming it is expected to become more frequent in the future. An even worse heat wave in Moscow (latitude 56°N, so even closer to the north pole than Paris) in 2010 killed over 10,000 people just in that one city. The cause was both high temperature and air pollution from raging wildfires in the surrounding areas. For more on this incident, see this URL: http://www.wunderground.com/blog/JeffMasters/over-15000-likely-dead-in-russian-heat-wave-asian-monsoon-floods-kil. Increased wildfires are one more anticipated effect of global warming, and this explanation has been cited to explain the unusually profuse fires in Colorado in the US and in Australia in recent years. It appears we are already living with global temperatures that verge on being too high for assured human survival.

Humans have only existed during the past few hundred thousand years, and in that time the global average surface temperature has been either about what it is now or as much as eight degrees (C) colder. It appears that this range may well be just about the maximum possible range of temperatures in which people can live without using some advanced technical measures such as air conditioning or space suits to shield them from the outside air temperature.

we run out of cool enough places to live. Eventually only the extreme polar regions and perhaps the highest elevations of the Alps, Andes, and Himalayas would be cool enough, and those high places may not have enough oxygen.

Where We've Been Getting our "Extra" Energy

As I mentioned above, like every other animal, we get some of our energy from metabolizing the food we eat. But we get a lot more energy from other sources.

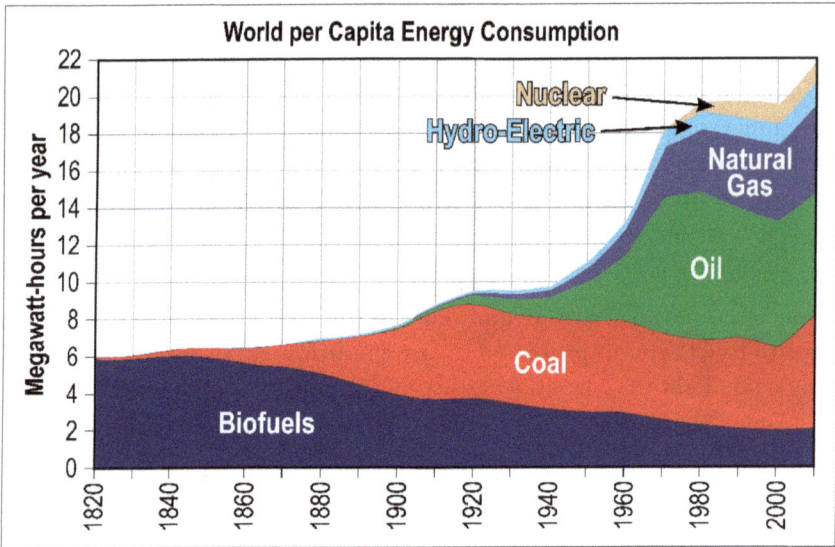

This figure shows how much we have gotten, per person (as a global average), from various sources in the past couple of centuries. Only the top two layers on this chart (nuclear and hydro-electric) are "green" sources. The amount of energy we've so far gotten from wind or solar power is, on this scale, insignificant.[12]

To summarize: This is the last of the three steps by which we've created our very big problem. We've used a whole lot of energy per person, there are lots of us, and we have gotten almost all of that energy from burning stuff. That has meant that we've dumped so much carbon dioxide into the atmosphere since the dawn of the industrial age that the level of CO_2 in the atmosphere now is higher than it has been any time in the past 40 million years. And we've done that really quickly—in less than 200 years we've raised the

---

[12] This chart is redrawn from one I found at this URL:
http://ourfiniteworld.com/2012/03/12/world-energy-consumption-since-1820-in-charts/
which is based on Vaclav Smil estimates from Energy Transitions: History, Requirements and Prospects together with BP Statistical Data for 1965 and subsequent, using world population estimates by Angus Maddison.

atmospheric level of carbon dioxide from about 280 parts per million to over 400ppm. Prior to about 1820 that level was very near 280ppm for all of the past 10,000 years, and it was never lower than about 180ppm in the past 65 million years, nor higher than about 300ppm in the past 3 million years.

Contrast the speed with which that level has changed lately with the roughly ten thousand years it took for that level to rise to 280ppm from the approximately 180ppm level that existed at the depth of the most recent ice age. We've raised that level more in 200 years than nature did in 10,000 years (between 20,000 years ago and 10,000 years ago).

Even if we immediately stop putting any more CO_2 into the atmosphere, left to nature, most climate scientists think the CO_2 level will take many centuries to decline significantly on its own.

How Can We Solve Our Problem?

What we need is a "detour" from this dangerous road we are on. Specifically, one that not only promises to let us avoid the disaster, but one that people can be induced to follow.

I know of three broad groups of people who have proposed some sort of "detour." The first group is focused on reducing our use of energy by conservation, increased efficiency, or simply by learning to do with less. In a way, the people in this group want to see us go back in time to when man lived a much less energy-intensive lifestyle.

The second type of proposed "detour" assumes we can, and now must, switch our source of energy away from burning fossil fuels to some more renewable, and preferably "greener" source. Some of these folks think just choosing to burn more renewable materials, like biofuels, will be sufficient. Others insist we must stop burning stuff altogether, and instead advocate for relying on nuclear, geothermal, hydroelectric, wind, and/or solar power.

The third type of "detour" suggests that we can keep on doing what we are doing about getting energy, but just mitigate the resulting rise in greenhouse gases—perhaps by a carbon tax to encourage the second group's notion of switching energy sources, or maybe by some form of "geo-engineering" that will either remove excess carbon dioxide from the atmosphere, or reducing the amount of sunlight arriving at the ground, thus keeping the Earth from overheating.

Unfortunately, none of these proposed "detours" has so far attracted a sufficient following to make it seem that it will end up being a likely way for us to stave off the impending mass-extinction disaster. And all of them, if implemented as their proponents suggest, amount merely to slowing down humanity's trip off the climate cliff.

The Problem with Detour #1

Conserving energy and increasing the efficiency with which we use energy are unquestionably good things to do...if they don't cost us anything. Unfortunately, they often do cost us—and often that cost is more than people are willing to pay. Furthermore, simply doing these things only minimizes the amount of energy we use, which doesn't truly fix our problem. Even with the best conservation and energy-efficiency practices we will use a lot of energy. Enough energy that we may well still drive ourselves to extinction if nothing else changes.

How about simply resolving to use less energy? That, for many reasons, is no solution. Perhaps the most important problem with that approach is its relation to our related problem of a currently scarily-fast-growing world population.

The amount of energy used by a few people and where they get that energy wouldn't have any appreciable impact on our global climate. The reason we have a real problem now is that (a) on average we use a lot of energy and almost all of it comes from burning stuff, and (b) there are so very many of us. So the total production of carbon dioxide and other greenhouse gases worldwide is now large enough to overwhelm other influences on our global climate—and do so in ways that could easily lead to our imminent demise.

An ever-growing population is a threat in another way as well. There is only so much land, so much potable water, and so much of every other resource we depend upon. At some point an ever-growing population will outstrip that supply of crucial resources. Some argue we have already passed that point, while others say we are merely getting close to the limit.

Reducing fertility is the only way, aside from war or famine or other ways that increase the death rate, to reign in the growth of population. And the only proven way to accomplish voluntarily reduced fertility is to provide people with *more* energy per person and at a cost they can afford. Only then, when they are able to lead more prosperous (which is to say, more energy-intensive) lives, will they voluntarily reduce their fertility. And once they do, the world population will at least stabilize and very possibly will begin to decline. For a discussion of the new problems that can cause, see the recent New Yorker or Time magazine articles by Elizabeth Kolbert and Hannah

Beech.[13] Or look at my companion book, *Multiple Population Problems and a Surprisingly Graceful Solution* (see footnote on page 6) which also talks about how we can solve all of these population problems using the same new infrastructure on which this book is focused.

The Problem with Detour #2

One sub-group of the people proposing we switch our energy source to something greener say we simply need to switch from burning fossil fuels to burning one or another sort of biofuel. Some of them propose using old cooking oil. Unfortunately, there isn't enough of that to make much of a difference.

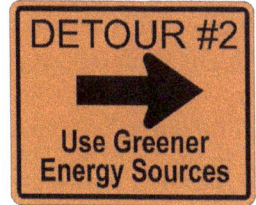

Another sub-group proposes that we switch to ethanol made from sugar cane, corn, or switch grass. A major problem with using corn for this purpose, as is now being done in the United States, is that it diverts corn that would have been used as food for people or animals to use as fuel instead. This can cause famine, or at the least, large increases in the price of all manner of food stuffs. And worse yet, the total "carbon footprint" (including the fertilizer and tractors used in the growing of the corn, plus the processing of it into ethanol and then what happens when you burn it) for ethanol as fuel is as bad or worse than the fossil fuels it replaces.

Further, when you raise any crop to burn as biofuel you are using up more of our limited supply of fresh water, and more of our land.

Both of these sub-groups are worried more about running out of fossil fuels than about the pollution of the atmosphere. They're right that we have literally been burning our way through a "savings account" of fossil fuels that has been many millions of years in the making, and doing those withdrawals with such rapidity that surely we will run out of all the saved up fossil fuels some time fairly soon.

My concern is that we may kill ourselves first. So I want to direct their attention away from the limited supply of fossil fuels and to the danger posed by the burning of *any* organic matter for energy.

Burning stuff (any organic stuff) to get energy is the crux of our problem. It always produces carbon dioxide, or methane, or oxides of nitrogen (or all of these), and all of those are greenhouse gases (GHGs).

---

[13] "Head Count," by Elizabeth Kolbert in the 21 Oct 2013 issue of the New Yorker, pp 96-99 and "The One-Child Reversal," by Hannah Beech in the 23 November 2013 issue of Time, pp 36-39.

The rest of the folks this group (those who are focused on how much energy we use and where we get it) say we should switch to some truly "green" alternative energy source. Some push for using hydropower, or geothermal energy. That would be helpful, if only we had enough places on the planet to use those processes. But we don't.

Still others propose switching to wind or solar power. They say we can overcome the fact that these sources aren't available 24/7 by finding ways to store energy between the times the wind blows or the sun is shining. That might be done—at some, as yet unclear, additional cost—but so far it doesn't appear at all certain that we could scale up these processes enough to meet fully our collective need for energy.

Finally, in this group are those who suggest that we should switch to nuclear power. I'll not dwell here on the problems with this idea. Some nations (in particular, France) have embraced this approach. Others have tried it and then rejected it. And in the largest energy markets (China and the US) there are some nuclear power plants, but far from enough to meet the bulk of their energy demand and it doesn't seem likely that enough of those plants could be built soon enough.

The real problem with *any* of these alternative energy sources turns out to be cost. They *all* cost more than just continuing what we are doing already. Every year the US Dept. of Energy publishes a document that, among other things, lists the cost to build and operate new utility scale power plants using various technologies. In the chart on the next page you can see the latest data they have published.

What you see in this chart[14] is the cost per Megawatt-hour for all of the generated energy a given type of power plant will generate over its entire life, including all the construction, operating, and maintenance costs, plus the fuel costs—all expressed in current dollars. The "error bars" on each bar in this chart show the range of costs they estimate for the corresponding technology for various locations within the United States.

The red bars are the "dirtiest" technologies. (Burning coal not only puts out more CO_2 per MWhr generated than burning natural gas, it also puts out fine particulate matter, sulfur, and mercury.) The orange bars are other technologies that involve burning fossil fuels, and the tan bar is for burning biofuels.

All of these put carbon dioxide into the atmosphere—except that those marked CCS. This indicates Carbon Capture and Sequestration—a technology

---

[14] US DOE Annual Energy Outlook 2013 with Projections to 2040.pdf; current data may be found at http://www.eia.gov/forecasts/aeo/er/electricity_generation.cfm.

that is supposed to remove (some, most, or all?) of the carbon in the exhaust from the power plant and store it safely somewhere (forever?). That technology has yet to be perfected and the manner and location for permanent storage is yet to be developed.

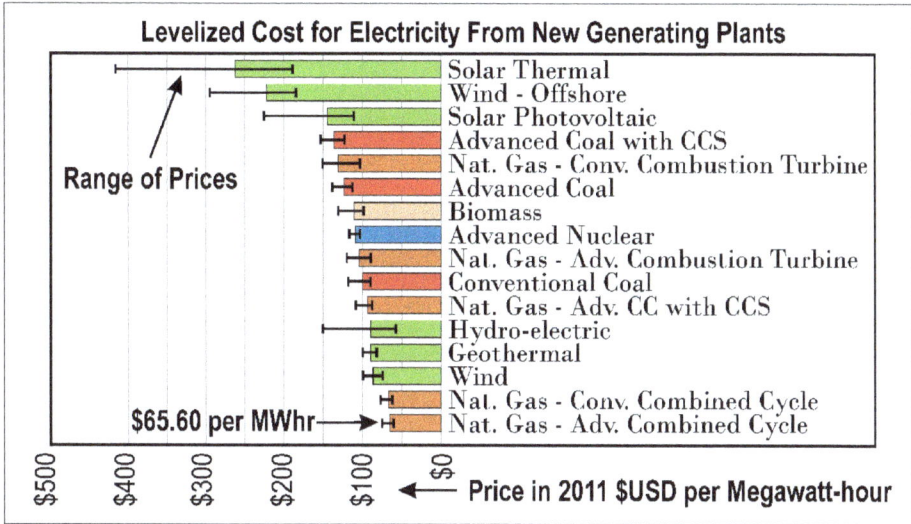

Levelized Cost for Electricity From New Generating Plants

The blue bar is for nuclear power, assuming someone figures out how to get around the various political and practical problems this technology presents. If that can be done, this would also qualify as a "green" energy source in that, apart from the construction of those plants (which takes a lot of cement—whose creation releases large amounts of CO_2), this technology at least doesn't produce any greenhouse gases during its normal operation.

The green bars are the "green" alternative energy technologies that this first group of folks are really focused on. But notice, please, that all of them—indeed *all* of these bars above the bottom one—are technologies that cost more than our present best way of getting energy by burning natural gas (without CCS).

That one point—cost—is the one on which most of the calls for switching to alternative energy sources founder. And with good reason. Money is tight. People don't want to pay more, especially for some future benefit that they cannot yet see is going to be essential.

The Problem with Detour #3

This group of proposed solutions include some who propose assessing a carbon tax, the proceeds of which would be rebated to citizens. This would put in place a market force against those energy sources that produce lots of carbon dioxide, and in favor of the "green" sources that do not. The rebates would compensate average folk who might have to continue using "dirty" energy sources for the increase in cost they would incur as the producers pass along the added carbon tax cost.

This might, over time, lower the amount of fossil fuel (and biofuel) burning to get energy, but it wouldn't, by itself, suffice to stop that sort of burning entirely. So this is, by itself, an incomplete mitigation.

Something more will be needed. The rest of this group of proposed solutions fall into the category called geo-engineering.

In an article titled, "Prozac for the Planet,"[15] Christopher Cokinos reviews the history of geoengineering proposals from the earliest suggestions almost a century ago to the most recent ones proposed in the last decade. He divides these proposals into two general type: Carbon Dioxide Removal (abbreviated as CDR) and Solar Radiation Management (SRM).

The first aims to undo the result of so much fossil fuel burning by taking excess CO_2 out of the atmosphere. That would lower the trapping of solar energy by the atmosphere, and thus cool the planet. The second approach cools the planet by reducing the amount of sunlight that enters the lower reaches of the atmosphere. Most of these proposals suggest doing that by raising the albedo (reflectivity) of the planet. All of the proponents of any of this category of "detour" start with the understanding that by pumping lots of carbon dioxide into the atmosphere we have increased the greenhouse effect, and that if this is left unchecked, it will cause the Earth's temperature to rise to the point that many species (possibly including ours) will perish. They all focus on how we might forestall that either by removing that excess carbon dioxide from the atmosphere, or by reducing the amount of sunlight that comes into the Earth. If we can do enough of either of these the Earth's temperature could be kept in check.

The source of much of his narrative was a conference held at Asilomar in California in March of 2010. It was attended by many scientists and others

---

[15] Christopher Cokinos, "Prozac for the Planet," American Scientist, Volume 79, Number 4 (Autumn 2010), pp 20-33.

from around the world, and the attendees discussed in depth all of the proposed means of geoengineering now considered as viable candidates.

Professor Cokinos' conclusion regarding CDR approaches is that there are some that probably would work over a period of many decades and at a large cost. But they are only a means to mitigate a portion of the on-going assault on our atmosphere by the burning of fossil fuels.

His conclusion about the present or proposed SRM approaches is that they may do the job, in the sense of reducing sunlight enough to hold the global climate more-or-less where it is now, but they will very likely introduce a number of not-so-pleasant changes to the Earth. Things like no more blue skies in the day time and no visible stars at night, uneven changes to climate with little in the way of obvious controls to smooth out those changes, and these approaches all require an unlimited duration commitment of money and effort to keep the benefits coming.

An aside: Many people talk about "Saving the Earth." We don't need to save the Earth, nor could we. It has been around for 4.5 billion years and is likely to remain much as it is now for at least another 5 billion years. What these people really want is to "Save the Holocene epoch." (That's what geologists call the current pleasant global climate which has only been around for the past ten or eleven thousand years.)

Humans have existed for longer than the Holocene, with the earliest humans appearing a few hundred thousand years ago. Since then we've lived through the latest of at least a dozen ice ages (which were separated by relatively brief pleasantly warm periods) before the planet finally started warming back up about 20,000 years ago and finally reached something like the present global average surface temperature at the start of the Holocene roughly 10,000 years ago— about when our civilization first took off.

We have records that let us infer the temperature of the Earth far back in time before there were humans, and those records show that between 10 and 60 million years ago, for example, the Earth was much warmer than it is now. So we know the *Earth* can survive getting warmer, but there is no assurance that *we* could survive that.

Overall, he concludes that it may be necessary to do something to control global climate change, and do it soon. But the best solution would be to stop burning fossil fuels—and that, he notes, can only be done if we find another way to generate all the energy we need and want.

None of the geoengineering proposals to date can be implemented without substantial expense, and none offers a way to generate more energy and thus to reduce the need to burn fossil fuels.

Also none of the SRM approaches does anything to mitigate the ever-increasing acidity of the oceans. When we pump CO_2 into the atmosphere, almost half of it gets dissolved in the oceans. And they are already getting so acidic that shellfish are having problems building their shells. Once these animals die, all the species that depend on them for food will also die. And so on, up the food chain. Similar things are happening to disrupt the food chain on land, such as increased temperatures making it infeasible to raise corn or wheat where we have traditionally done that.

When too much of our food disappears, moving to higher latitudes or altitudes won't be a useful option.

So these sorts of geo-engineering scheme aren't really solutions. They don't provide us with more energy, and they don't address world population growth. They at best mitigate a part of the problem of global climate change. That is sort of like putting a band-aid over a break in your skin after you take a bad fall, when your real problem is the broken bones underneath. These sorts of change wouldn't prevent the mass extinction event. Instead, they would merely postpone it for awhile.

So I would summarize all of the "detours" I have discussed above as really *grim* possibilities. None is going to do the whole job, and all are unpleasant in at least how much they cost—while several also are grim in the way they would destroy our quality of life.

Part Two
A Real Solution Through Building an Infrastructure

Enough with the Grim Detours.
How About A Real & Delightful Detour?

What we really need is another sort of detour altogether. One that addresses all the dimensions of our problem at once (Our need for energy, our need to control our world population, and our need to control global climate change). One that doesn't require sacrifice. One that promises a better experience, sort of like when a superhighway opens up next to the old and now broken road that you've been traveling.

I'd call any such "detour" a delightful one (as opposed to the grim detours I described above). And that exactly describes the results from the new infrastructures I am proposing we build.

They will completely remove the problem of global climate change, plus they will provide us with all the energy we are likely to want now and into the future, and in the process also will enable us to address our world population problems the right way. At the same time they will enhance our lives in many other almost unimaginable ways. Even better, instead of costing a lot more than our present course, once they are built, they'll lower the cost for energy and transportation dramatically. (Cost will be reduced by at least a factor of ten for energy and by a factor of as much as one hundred for transportation—

and the transportation will be speeded up by a factor of roughly one hundred as well.) This speeding up and lowering of the cost of transportation will cause our world economy to grow explosively.

Once we build these new infrastructures we will be able to stop burning stuff to get energy. Completely stop it. In addition to providing a truly green source of all the energy we could possibly use, they also directly address global climate change by controlling the amount of sunlight reaching the planet. And, as a bonus, they will permit us to exercise a heretofore unimaginable level of control over regional weather. (Beyond all that, they have a number of really wonderful bonus features—including mass market space travel. I'll briefly describe just a few of these bonus features in a later section.)

For the most part, what I am proposing uses already proven technologies. It simply puts them in a new context where their performance is radically better. Building that new context, on the other hand, will be a major engineering challenge.

I cannot absolutely assure you that we *can* build what I am proposing. But, after checking with a number of experts, I have been unable to find anyone who can cite a valid reason why it shouldn't be possible to build it. And I've recently found a group of people who have been exploring a closely related problem. The results they've come up with (which I will describe in the section titled "The Tall Tower Project" starting on page 43) make my proposal seem very plausible.

It will be a challenge. It will be the largest engineering project ever undertaken by mankind. And surely there will be a number of unanticipated problems to be solved in creating it. But we've done that sort of thing a number of times in the past. We just need to know how crucial it is that we make the attempt, and then put our resources behind solving any engineering problems that arise along the way.

It will likely cost a lot to build this new infrastructure, and it will also require solving some serious political problems, but the resulting growth in economic activity will be such that even a very modest tax on that activity will easily repay the initial investment many times over. Finding out whether or not we can build it can be done with only a modest initial outlay of resources.

And, once we know we can build it, committing to doing so will create the largest jobs program and the greatest economic boom in history, plus transforming our lives in numerous and, I assert, mostly quite enjoyable and beneficial ways.

A Brief Description of the New Infrastructures

Enough of the reasons why I think this is both crucial and wonderful. Now it is time to tell you exactly what this new infrastructure is.

There are four essential components to the overall infrastructure, plus a number of optional extras that could usefully be added to it. The first one provides the novel context for all the others. The second is a context-appropriate version of solar electric power generation and distribution. The third is a context-appropriate version of a magnetic levitation train system. The fourth portion uses very simple and proven technology to do a task that simply cannot be done at all without this novel context, and its purpose is to control our global climate and, potentially, each region's weather as well.

And once we have those four components in place, there are many additional features that can be added at very little cost to do jobs that are now either hugely expensive or completely impractical.

Component #1 "The Platform"

To visualize the context I want you to think about a trestle or viaduct. Originally these were invented by the Assyrians about 2600 years ago as a way to carry water from mountain rivers to cities. They needed a way to build a channel that could let the water flow by gravity along a path that had to cross some valleys. The Romans adopted this technology and used it both for that purpose and also to let horse-drawn carriages travel more easily across a very unlevel terrain.

| 19th century trestle | 20th century trestle |

We use this technology daily to carry trains and cars across both smaller and larger valleys without their having to go down one side and up the other.

There are only two salient features to a trestle. One is that its top is at an essentially constant altitude. The other is that its very existence is of no importance except for what can be put on top of it!

Now think about a really high trestle. One whose top is at an altitude of 100km (about 60 miles) above sea level. A system of such trestles interconnecting all the major metropolises on Earth is one way to describe the first, context-creating portion of my proposed new infrastructure.

And, because it only is important for what it supports, I call it **The Platform**. Here is a conceptual drawing that shows one Platform trestle segment from Los Angeles to Paris as a thin magenta line. This view is from a vantage point in space about 7,000km above the North Atlantic. In actuality, the way The Platform is likely to be built would make it nearly invisible from any distance of more than a few hundred meters away.

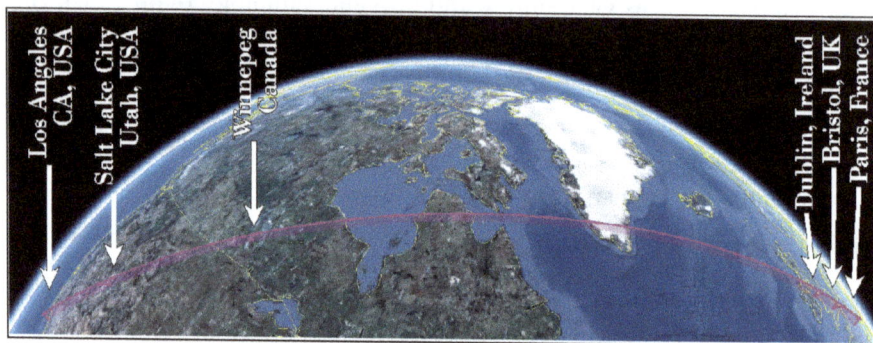

The light magenta line represents the top of The Platform. I've given a light shading from the magenta line (which is at an altitude of 100km) down to the surface of the Earth to suggest The Platform trestle's equally invisible support structures.

The significance of the various cities indicated on this diagram (and the curving magenta lines connecting those cities to the top of The Platform) will be described in the section titled "One Trestle Serving Multiple Cities" starting on page 87. The Platform trestle segments are really long (in this case about 9,000 kilometers) and really high so they also are likely to be very wide at their base, where they are supported on the ground—perhaps some 20-30 kilometers wide. Given that overall size, it seems reasonable to specify that the top should be made at least five kilometers wide to support all of the other component infrastructures that it will be supporting.

I know this sounds really audacious, if not actually absurd. The tallest structure that has been built by humans so far is just slightly less than one kilometer tall. Yet I am proposing that we build not one, but many structures that are a hundred times taller than that. Seriously.

When sharing this idea with one friend, he told me about something Albert Einstein reputedly said: *"If at first the idea is not absurd, then there is no hope for it."* Well, by that criterion, this may be our best hope yet!

Is it Possible to Build Something That Big?

As I said earlier, none of the experts I've spoken to (and there have been quite a few) can cite any good reason why this is impossible. All of them agree that it would be unprecedented, but then that is the essence of all really revolutionary ideas.

I started by asking this question: "Is there a theoretical limit to the height of a self-supporting structure built on the surface of the Earth?" Engineers quickly explained to me that this is simply not an engineering sort of question. They said, "If you come to me with a business need and a budget and then ask me if I can build something, *that* is an engineering question."

So perhaps the best way to describe our present situation is that we know how to build structures up to almost one kilometer tall. And, in fact, the only reason so far to build even these structures has been bragging rights to owning the world's (or your region's) tallest building. Most of the really tall skyscrapers built to date are not totally occupied. No one knows how to use them fully effectively, and clearly even any moderately taller towers would be even less-well used.

Recently I gave a draft copy of this book to Gregory Benford. He is a physicist on the faculty at the University of California, Irvine, and a famous science fiction author. His reaction was to tell me that I needed more numbers for this section of my book. And, thankfully, he directed me to a group that has been studying the possibility of building a really tall tower for the past few years.

The Tall Tower Project

That group came together at a web site created by the Arizona State University. They called that web site Project Hieroglyph. Its purpose is to be a place where writers can propose "outlandish" ideas and get scientists and engineers to give them a reality check. Or anyone could propose some idea and have the others who frequent the site give them feedback on it.

The Tall Tower project[16] began with a couple of questions by Neal Stephenson that were rather close to the one I'd been asking. His questions were: Just how tall a tower could one build using today's tall structure building technology? And once you know the answer to that question, what useful thing could one do with it?

My question, in contrast, didn't limit the technology to what we know how to use today, and I had in mind both a specific height and purpose for

---

[16] For a summary go to http://hieroglyph.asu.edu/project/the-tall-tower/

building these tall structures. Furthermore, my structures were to be trestles, rather than simply an isolated tower. (Also, they were to have some other special properties that I'll discuss in the section titled, "The Necessary Special Features of The Platform" starting on page 89.)

Over the next couple of years, with input from a NASA physicist, a structural engineer, and some others the Tall Tower group came up with an answer: We know pretty well what it would take to build a steel tower up to at least 15km and perhaps as much as 20km tall that could support a several hundred ton payload. Their thought was that this might be a useful platform from which to launch space rockets more efficiently than our present strategy of launching them from ground level.

Studying what they had done gave me several important insights. Here is some of what I learned:

The first fact one can use to get a very rough estimate of the tallest structure that could be built is the strength to density ratio of the structural material you plan to use. And that ratio, when the strength and density of the material are expressed in the appropriate units, is a length. For structural steel that length is about 15km.

One way to see the significance of this length is to imagine that you are supporting some specified load weight. You will need to use enough area of steel to keep the load from crushing the support material. Now imagine that the load has to be held up at a height and you use the same support members (same diameter rods, for example) and just lengthen them. Now the bottom portion of those rods will have to support the load plus the weight of the support structure. If you keep raising the load, eventually the weight of the support structure (plus the load) will crush the support material near its bottom. In the limiting case of a small load, the height at which this happens is just the strength to density ratio.

You can go somewhat higher by making the bottom portion of the structure broader than the higher parts. Think of the Eiffel tower. Its base is more than ten times wider than the top section of the tower.

A more complete analysis suggests that using steel, with a suitable safety factor, it should be possible to build a tower that wouldn't fall down under its own weight (plus a several hundred ton load on top) up to 20km tall.

Building something that won't fall under the force of gravity isn't enough. You have also to make sure that it won't be bent or broken. This is the sort of problem that structural engineers have to face on every structure they design, and there are many well-known solutions, such as replacing individual structural members with a truss arrangement, or adding cross-bracing elements.

Furthermore, you have to be sure that your tall structure won't be knocked over by whatever may buffet it once it is up there. In the Tall Tower project, it turned out that the jet stream winds might present the greatest challenge for such a tower.

Jet Stream Winds

All weather on the Earth is caused by the interaction of sunlight and the rotation of the planet. The sun shines most intensely on the tropics, which warms the air there more than at the poles. This means that the hot air in the tropics will rise and flow out both north and south toward the poles, where it will be cooled and fall down the ground once more. The cooler polar air will flow near the ground in the opposite direction to replace the rising hot air near the equator. However, the Earth turns constantly, and this changes things.

Since the Earth is always turning there are gyroscopic forces pushing sideways on the air as it moves either toward the poles or back toward the equator. For a planet of the Earth's size, this naturally breaks up the overall pattern into six cells, three in the northern hemisphere and three in the southern hemisphere. The Hadley Cell is the name given to the two cells that extend either north or south from the equator to roughly 30 degrees N or S latitude. The Ferrel Cell is the name given to the next pair of cells, extending roughly from 30 to 60 degrees N or S. And the Polar Cell is the name give to the remaining cells (from where the adjacent Ferrel Cell ends to the north or south pole). The subtropical jet stream is a strong river of air flowing generally west to east near the top of the troposphere[17] along the boundary between the Hadley and Ferrel cells. The polar jet stream is a similar strong river of air flowing near the top of the troposphere where the Ferrel cell meets the polar cell.

Generally the air is warmest in the Hadley cells (the ones nearest to the equator), and coldest in the polar cells. This temperature difference is greatest in the summer which causes the jet streams in whichever hemisphere that is currently experiencing summer to move away from the equator toward to pole during the late spring and summer, and then reverse course and move back toward the equator in the late fall and winter. (Naturally, the jet streams in the other hemisphere do a similar dance but six months later, since summer in the Northern Hemisphere is winter in the Southern Hemisphere.)

---

[17] The troposphere is the lowest layer of the Earth's atmosphere. The next higher layer is the stratosphere. The thickness of the troposphere varies from about 20km (12 miles) to as little as 7km (4.3 miles) in the polar regions. It is the region where almost all of the Earth's "weather" happens.

The strongest portions of each jet stream is only a few kilometers in height but may be up to a few hundred kilometers in width (north-south) with the wind speed rising toward the center of that region to a maximum that depends on the amount by which the air temperature to the north of it differs from that to the south of it. Typically this peak wind speed will be in the range from about 100 to 400km/hr (about 60 to 250mph).

The jet streams meander as they go around the globe, and these meanders move in the same direction (west to east) as the wind inside the jet stream, but at a much slower rate. The polar jet streams are more compact and more intense than the subtropical ones. These polar jet streams are the ones most often spoken of, and they are the ones commonly used by airplanes seeking an advantage of a tail wind when flying eastward.

These polar jet streams are generally located between about 9 and 14 kilometers above mean sea level. So the force of these winds on the Tall Tower was a serious concern to the group planning how it might be built, once they realized they were talking about a tower that might be more than 10km tall.

To get an idea of how much force these jet stream winds can exert, you need to know that the force of wind on a stationary object is equal to its speed times the density of the air times some number expressing how much that object impedes the flow of air past it. The air at the altitude of the polar jet stream is roughly one-quarter of the density of air at sea level.

The speed of the wind in the jet stream maxes out at roughly the same speed as the most intense gusts in the next-to-strongest tornados (F4 on the Fujita-Pearson scale) [where damage is described as "devastating"], but because the air is so much less dense up there, the force exerted by even the fastest jet stream wind is around that of the strongest gusts of an F0 or F1 tornado [where damage is described as "light" to "moderate"]. (The force of a windblown object can be much more concentrated, which is how even an F3 tornado may be able to drive a bit of straw through a telephone pole, to cite one example.)

A structural engineer at Arizona State University, and one of the key contributors to the Tall Tower project, Keith Hjelmstad, did the calculations and came up with a tentative design for a tower that would be 15km tall and strong enough to withstand the jet stream winds. This wasn't, he emphasized, a final detailed design, but rather a model showing that it should be possible to create an actual buildable design for such a tower.

What the Tall Tower Project Teaches Us

In my conversations with a couple of the people who worked on that Tall Tower plan they expressed the belief that if one used newer materials with a higher strength-to-density ratio, it should be possible to build a tower as tall as the trestles I am proposing we build. It doesn't appear that the jet stream would be any more of a problem for these taller structures, since the higher you go the lower the air density, and so once you are above the jet stream both the wind speed and certainly the wind pressure will decline the higher you go.

These alternative materials may have a much higher strength to density ratio, but most such materials have an absolute strength that is far less than that of steel. This could mean that one might choose to build the trestle in two or three sections. The very topmost portion (perhaps only a few meters high) could be built out of some very strong material as a very strong and very rigid platform—one might use steel for this portion. That platform top layer would support the many things that will be placed on top of it, and spread out their gravity loads across the entire width of this top section. At the bottom of this section of the trestle this load would be transferred to some much broader support columns made from the alternate weaker but very much less dense material. This section of The Platform trestle could extend almost down to the ground. The bottom section (perhaps only the lowest few hundred meters of the trestle) could once again be made from steel or some other stronger but much denser material to minimize the area of the points of contact of the trestle with the ground.

Knowing that it should be possible to build the trestles that tall and that the jet winds won't be too much of a problem doesn't mean there won't be other problems to be faced and surmounted. NASA discovered quite awhile ago some possible problems based on their research and experience. One example is that above about 25km above sea level the oxygen in the air absorbs enough ultraviolet light from the sun to cause the oxygen molecules (O_2) to break apart into isolated oxygen atoms. This atomic oxygen is very reactive and can attack many substances, essentially burning (which, in the case of iron, we call rusting) the surface. So one must choose materials that are relatively impervious to that attack for the portions of the trestle construction that are in that region—or else one must coat those surfaces with some protective coating. This is mostly a problem at altitudes that are not too far above 25km since the amount of air, and thus the amount of oxygen in it keeps going down the higher you go.

Still, this can be a serious problem even for orbiting space craft in low Earth orbit (at altitudes around 150km to 400km above sea level). It is mainly a problem for the leading edges of a space craft in these orbits because they are moving through this thin air very rapidly, and therefore their leading edge runs into many more oxygen atoms each second than the number that would land on a stationary surface.

Stationary structures, like the trestles I'm proposing, would not experience nearly as much of a problem with atomic oxygen as orbiting space craft do. But some portions of it might be affected enough that protective measures should be taken there. This is just one example of a possible problem the builders of these trestles will have to consider. Very likely there will be at least a few others as well.

Some Advanced Materials

I mentioned that we may have to build The Platform out of some advanced material that has a higher strength to weight ratio than steel. Already there are several such materials known and in common use.

Almost two decades ago a NASA physicist, Geoffrey Landis, published a paper proposing that a very tall tower might be useful for launching space craft. [18] (He later on was one of the contributors to the Tall Tower Project I mentioned above, and one of my sources for this book.)

In it he cited four materials[19] that were available on the market at that time and that all have a compressive strength to density ratio at least seven times higher than steel. So, if we can build at least a 15km tall tower using steel, we surely should be able to build a 100km tall trestle using one of those materials, if the strength of the material is the principal limiting factor. And there are now many other materials that may prove even more suitable. (I'll mention one in just a moment.) One of the materials he mentioned is what has been used for decades now to make "carbon fiber" tennis rackets, violins, and other things...and more recently certain airplane bodies. The strength to density ratio for that material is about 108km.

Scientists have recently learned enough about how materials acquire properties such as strength and lightness to be able to design materials that have never existed in nature—ones that have radically different properties from anything we've ever had to work with before.

These insights about how materials are structured at the nanoscale and how rearranging them can modify their properties is an example of a modern materials revolution: The creation of "meta-materials." I'll give you just a few examples here.

One of the most astonishing (and perhaps relevant) of these new materials are the aerogels. These are very lightweight materials that nonetheless have

---

[18] The Tsiolkovski Tower Reexamined, Journal of the British Interplanetary Society, Vol. 52, 175-180 (1999). First presented as paper IAF-95-V.4.07 at the 46th International Astronautics Federation Congress, Oslo, Norway, 2-6 October 1995.

[19] Those materials are SiC/Epoxy, Graphite/Epoxy, B4C, and Boron/Epoxy.

substantial strength. Some have been built using silica (basically the stuff of which glass is made); others using carbon nanotubes. Here are some pictures showing samples of aerogels and demonstrating just how light they can be.[20]

Figure **A** shows a graphite aerogel that is super-light and yet quite strong resting on a blade of grass. Figure **B** shows a nearly invisible silica aerogel block (the wispy blue portion at the bottom) that weighs only two grams and yet is able to support a brick that weighs 2,500 grams. Figure **C** shows a thin slab of silica aerogel protecting some crayons from the heat of a torch (showing another special quality that this sort of material can have).

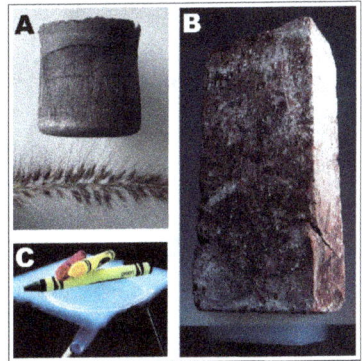

Another kind of aerogel (not shown here) is called a multi-walled carbon nanotube aerogel. This material can be squeezed down to less than one-tenth its size and, once you release the pressure on it, it will bounce right back to its original size.

Yet another variant on this idea is the cross-linked aerogel (referred to as an X-Aerogel). This is a material that is formed by coating an aerogel with a polymer material that will bond to the aerogel and form cross-linkages between the constituent parts of the aerogel. Doing this slightly increases the material's density (weight per unit volume), but it also makes it waterproof, more flexible and stiffer that the un-cross-linked version, and it in some cases it will dramatically increase the material's strength.

One sample X-Aerogel[21] discussed at the Aerogel.org website has an ultimate strength to density ratio that is roughly 16 times that of steel. This material weighs about 40% as much as an equivalent volume of air, so it has less than one ten-thousandth the density of steel. This suggests that its strength is somewhere around one five-hundredth that of steel. And thus you'd need about 500 times as much area of this material as you would of steel to support a given load. But that should be quite feasible with a structure the size of my proposed trestles. Recall that I suggested that they be at least five kilometers wide at the top. I would hope that the necessary material wouldn't fill this width, but would instead let all of the structure be built in a lacy

---

[20] A good resource about aerogels generally is at http://www.aerogel.org/.

[21] "Strong and Flexible Aerogels" http://www.aerogel.org/?p=1058, see the graph in the section "Applications for X-Aerogels" which is credited to Professor Leventis at the Missouri University of Science and Technology.

fashion so it won't block much sunlight, while still being amply strong to hold up the various components built on top of it. And if it were possible to build it using a silica X-Aerogel, it might be translucent and so would block even less sunlight.

A yet more recent innovation is the metallic microlattice. Made of nickel, which is a pretty heavy metal, this structure is so light it can rest on a dandelion seed head without damaging it. You can squeeze this material down to half its original size, at which point part of the material can easily be seen to have been crushed, but when you release it, it spontaneously un-crushes itself—quite unlike any macroscopic lattice structure built out of conventional materials. (An automobile whose body was made of this material might get crushed in a crash, and then bounce right back to its normal size and shape with hardly a dent!) This novel meta-material weighs only about three-quarters as much as an equivalent volume of air. But it isn't the record holder for light and strong materials. Another lab announced the creation of a graphene based micro-lattice material that is only one-sixth the density of air.

I'm not proposing that any of these particular materials be used in the construction of The Platform. I'm just pointing out that we are at the very beginning of a revolution in making materials to order with almost any set of properties we can imagine. Surely some of these new materials will be just what we need for building The Platform.

What About Building The Platform Over Oceans and Mountains?

This is another aspect of the project that I cannot cover here in any detail. You will find a substantial consideration of it in my longer book on this topic[22]. In summary, I have found a number of strategies that can be used in various situations to build The Platform without any workers having to work at altitudes more than about a kilometer above the local ground level, and ways in which it can be built across any topography from the deepest oceans to the highest mountains.

---

[22] Removing the Three Biggest Threats to Humanity, One Real Sustainable Solution to Our Energy Crisis, World Population Explosion, and Global Climate Change by John M. Goodman, Ph.D., forthcoming.

Another Very Large Structure Being Seriously Pursued

It may give you some comfort to know that there is one other similarly big idea out there that has been seriously pursued since at least 1969, and on which substantial progress has already been achieved. It is even larger in one way. This is the so-called "space elevator" project. Here is a drawing of a space elevator attached to an anchor point in Ecuador.[23]

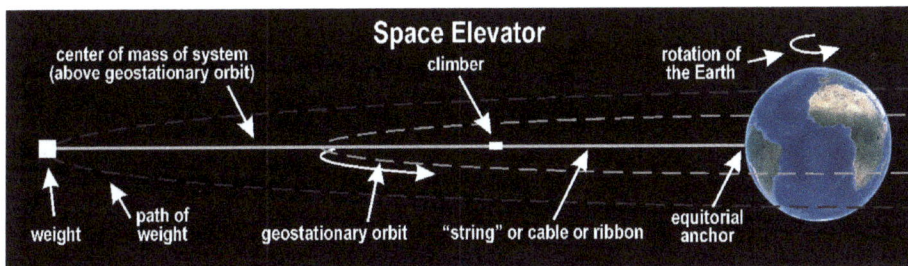

Space Elevator

Think of swinging a rock tied to a string in a circle around your head. Swing it fast enough and it will pull the string taut. Now visualize putting a very large and heavy object on the end of a really long "string" that is tied to the Earth somewhere on the equator. If the string is about 100,000 kilometers long (a full thousand times longer than the height of the structures I'm proposing) the heavy object at its end will be whirled around in a circle once each day as the Earth turns, and that will pull that string taut and keep the system stable.

Now you can imagine a small climber that grabs onto the "string" and climbs up from the surface of the Earth to some altitude where it can release a package it has carried up there.

If it climbs to an altitude of approximately 35,786 km (22,236 mi) above mean sea level, the climber will be traveling around the Earth at the right speed to be in orbit, so if the climber has carried up a satellite, it need merely let it go and the satellite will just stay where it is released. That is one way to launch a satellite into a geostationary orbit without using any rocket power at all. (And with minimal "station keeping" jets or some other strategy, that satellite can move around its orbit to whatever location best serves its intended uses.)

---

[23] You can read this history of the space elevator at this URL: http://www.star-tech-inc.com/id4.html, and you can get an update on current progress and future plans at http://www.isec.org/.

Carry the package to a higher altitude and when it is released that package will be launched on its way out of Earth's orbit and, perhaps, on its way to the Moon or Mars or beyond.

In order to keep the ribbon or cable taut the weight at its end needs to be far enough out from the center of the Earth that the center of mass of the entire space elevator system is above any plausible space vehicle release altitude, which is why the designers need to have a cable length of roughly 100,000 kilometers (or more.).

This sort of alternative space launch capability has received a lot of attention in the past several decades, with engineering conferences to present papers on various aspects of the technical challenges that must be surmounted to achieve its construction, plus some X-prize type contests to build proof-of-principle demonstrations of some of the key parts of this system. And, I remind you, this is a plan to build something *one thousand times taller* than the trestles I have in mind.

Researchers feel that they have figured out, in principle, what would be a strong enough material to make a practical space elevator. Now they just have to figure out how to make that material in industrial quantities.

There are two significant differences between the space elevator and what I am proposing. The first difference is that the space elevator uses a "string" (which isn't really a simple string, but is likely to look more like cable or ribbon) that needs to be very strong in tension, but doesn't have to have any strength in compression or bending. The trestles will likely involve materials that are strong in both tension and compression, and depending on how it is built, they may also have to have some substantial bending strength as well. Some of the properties of the latest aerogels and microlattice materials are looking quite promising, with still more—and possibly even better—materials likely to be discovered by the time we need them.

The second, and even more important difference is that the space elevator does only one job: Launching space vehicles. It doesn't provide a way to generate lots of energy, nor does it do any of the other things my proposed new infrastructure will do. So I only mention it here to point out that mine is not the first proposal for a really large structure that is being taken seriously.

And to distinguish mine from the space elevator, since my proposal solves many aspects of our problem at once.[24]

The Platform Could Be Built—And It Should Be Built

I cheerfully agree that building such high trestles will be a major engineering problem, and will require substantial research and perhaps some inventions before it can be completed. But that doesn't mean it can't be done. **And if you understood and were convinced by Part One of this book, you may now realize that we might have to do this if we are to continue living on Earth. That is our "business need" for building The Platform and the other infrastructures that I am proposing that we build on top of it, and I'm urging that we do so just as soon as we can.**

As to having the necessary budget, that is quite another matter. Since I don't know for sure *if*—let alone *how*—we can build it, I cannot say at this point how much it will cost. But if it is seen as the best way to save ourselves from extinction, I believe we *will* find a way to afford it. And, expensive as it is likely to be, I think it is very plausible that we can find the money to build it—especially since the history of transportation infrastructures suggests that it will be a very good investment.

I'll give you my thoughts briefly on just how we might fund it—and why it will turn out to have been a very good investment in purely financial terms—in the next section. You'll find more details on this in my longer book (see footnote on page 50). There I also speculate on just how we might do the construction, as well as what sort of design it should have.

Where Will the Money Come From to Build The Platform?

This is an excellent, and obviously important, question. It is helpful to divide up the overall project of building The Platform and the other infrastructures it will support into two phases. First is the research phase to prove that it can be built (and learn how to build it and therefore what it will cost). The second phase is the actual construction.

I think the first phase may be funded by a combination of grants by charitable organizations, wealthy individuals, and government agencies. It

---

[24] Other large alternative space vehicle launch systems have been proposed, with names like the Space Fountain, the Launch Loop, and the Space Cable. All of these have the same drawbacks as the Space Elevator; they only launch space craft. They do nothing for energy production (indeed, some of them use fantastic amounts of energy) and they do nothing for point-to-point terrestrial transport, let alone doing anything to help with controlling global climate and/or regional weather.

might best be organized as a series of contests, similar to the various X-prize competitions, with prizes to be paid to the teams that manage to achieve various milestones. That way, only the prize money is at risk by the funding organizations. Each team will have to find its own funding in the hopes that it will win the prize—and to gain the expertise to become a dominant bidder for the final construction contract.

Once the first phase is complete, the second phase funding can be arranged. At this point the cost to construct at least the initial segments of The Platform (and the associated solar power, transportation, and other infrastructures to be built on top of The Platform) can be estimated with some precision. And so can the benefits. With that information in hand, it may be possible to float a large, long-term bond issue to acquire the funds for the construction. The bond holders could be guaranteed a fixed interest rate as a minimum benefit, plus some small percentage tax on new economic activity enabled by this new infrastructure as a speculative additional payback.

I'll give you some of my personal speculations about how the second phase may play out in my Summary Comments section near the end of this book. Again, this is a topic I plan to cover in more detail in my longer book (see footnote on page 50).

I must urge that at this point those who put up the money for The Platform's construction should NOT expect to be repaid by a share of the revenue paid by the users of the infrastructures on top of The Platform. Instead, they should view The Platform as an enabling technology that will give such a boost to overall economic activity that the investors can easily, and very generously, be repaid by just a tiny tax on that increased economic activity. This puts the cost burden for The Platform on everyone in all the societies on the Earth—which is appropriate, as they are going to be the ones getting the benefits (both survival instead of extinction, and greatly expanded economic opportunity) from The Platform's existence.

It may well be that the initial construction cost of some or all of the other infrastructures that are built on top of The Platform would also be viewed as a part of that initial enabling investment, to be repaid along with The Platform's construction cost out of the general growth in economic activity that we can be sure will flow from these new infrastructures.[25]

---

[25] James Suroweiki, in an essay "Sputnikonomics" in The New Yorker (Feb 14, 2011) cited one study of the U.S. Interstate Highway system that reported a 35% annual rate of return on the investment needed to build the system. I suspect the return on investment on the Hyper-Speed Transport System would be at least that large over any substantial time period after it is fully constructed.

After all, everyone on the planet will benefit from our shift away from fossil fuels, as well as from the increase in economic activity. So the initial cost to create The Platform and the associated other infrastructures should be born in some manner by everyone.

That is the smart way to structure the funding if the major funder is a government that wants to grow its economy. If a major funder is a business that wants to grow its business, they can look to using a share of their gross profit on that new business to repay their investment. I expect the actual funding will come from some combination of governments and multinational corporations, plus perhaps some non-profit organizations and wealthy individuals.

The users of the infrastructures that operate on top of The Platform should expect to pay *only* the cost to operate and maintain those infrastructures (plus some profit for those who run those infrastructures). The cost of the enabling Platform should be paid by the overall society and/or by the major companies that benefit from the existence of these new infrastructures.

* * *

Again, I cannot assure you that we can build The Platform. But it seems very plausible to suppose that we can, and we can certainly do the necessary proof-of-principle tests for a rather small investment, and once that succeeds, the rest will be just an ordinary (albeit massive) engineering and manufacturing/construction effort.

There are many details about how The Platform must be built to do its job properly, most of which are required for the third infrastructure I'm going to describe, so I'll describe some of those details in a later section titled "The Necessary Special Features of The Platform" starting on page 89, in connection with my discussion of that other infrastructure component.

Part Three
Getting All Our Energy From the Sun Nearly For Free

Component #2 Energy Generation and Distribution or "The Global Electric Grid" (GEG)

This, and all of the other infrastructure components I'm going to describe utilize only well-established technologies. There will be no huge engineering challenges left to be surmounted, once The Platform has been built.

Generating Power from Sunlight Efficiently

We've known for a long time that we could easily get all the energy we could possibly use from the Sun—if only we could put solar panels outside the atmosphere and then somehow get that energy back to the ground safely and economically. Once we have The Platform built, we can mount the solar panels on top of The Platform and connect them with one another and deliver the electric energy they generate to the ground by superconducting cables running on top of The Platform and down to the ground wherever there is a demand for electricity. I suggest the name **Global Electric Grid** (or **GEG**, for short) for this generation facility and all of the interconnections it needs to deliver that power to the ground.

The physics of generating and transmitting electric power from solar panels is profoundly different when they are outside the atmosphere from what it is for solar panels down at ground level. The technology and engineering are mostly the same, but how they perform is radically different.

The primary reason solar panels would work so much better up there is that there is a whole lot more sunshine up there for the panels to work with. About six percent of the sunlight falling on the outside of Earth's atmosphere where it comes in straight down simply bounces off. A somewhat larger percentage bounces off when it comes in at an angle to straight down, and a much larger percentage bounces off where it comes in at nearly a grazing angle. Furthermore a good deal of the sunlight that enters the atmosphere is absorbed before it reaches the ground.

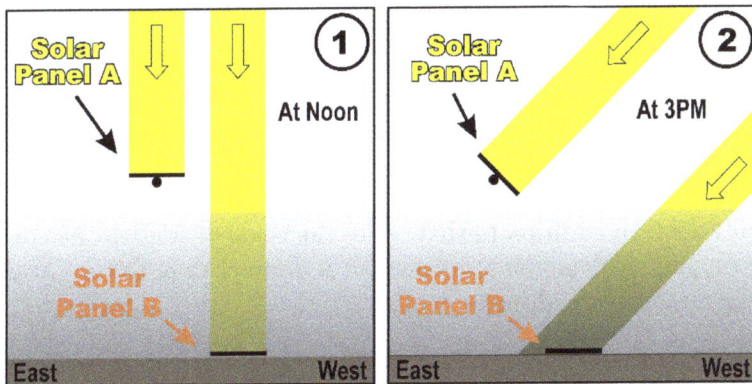

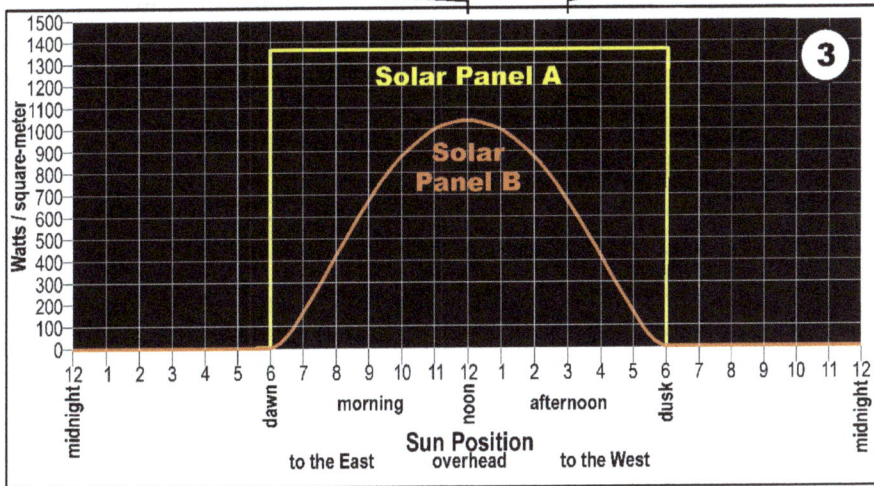

This figure represents what happens during the day for either of two very special days: the spring and fall equinoxes, and for only one location on the Earth. Look at the upper left boxed portion of this figure, labeled ①. Here the solar panel **B** on the ground is located precisely at the equator and it is in a fixed position, parallel to the ground. On those two days, and only on those two days, and only at this location, the sun travels directly over the solar panel at local noon. At that moment the sunlight hitting the panel has traveled the minimum possible distance through the atmosphere, and thus is the brightest that sunlight can ever be anywhere at ground level. At every other day in this location, or any day anywhere else the sun will not be directly overhead at noon, and thus the peak brightness for that day won't be as great as this. Solar panel **A** is located above the atmosphere and it rotates to always point at the sun. It always sees the same full solar brightness which is about 35% more than the peak brightness seen at ground level.

In the upper right boxed portion, labeled ②, you see the situation at 3PM in that same special location and on that same special day. Now the sunlight reaching the panel on the ground is diminished in two ways. First, the sunlight is now coming in at 45 degrees from straight down. That means that solar panel **B** is, from the point of view of the sun, tipped away 45 degrees. This means it will only intercept about 70% of the beam that it originally intercepted. (The beams of sunlight in this figure are the same width for both solar panels and at both times of day.) This first effect could be termed the geometric effect. In addition to that, since the beam of sunlight has to travel through roughly 40% more atmosphere, it is dimmer than the one at noon. This is the atmospheric absorption effect.

Now if you look at the lower boxed portion, labeled ③, you'll see a graph of how much sunlight energy is deposited on each solar panel at each hour of the day from midnight to midnight. During the hours of darkness, neither panel sees any sunlight. During the daylight hours solar panel **A**, which is always aimed at the sun and is always bathed in full strength sunlight, receives 1365 watts of solar radiation per square meter. During the daylight hours solar panel B at first sees only very weak early morning sunlight. As the day goes on, it intercepts a wider and wider portion of the incoming sunlight beam, and that beam gets more and more intense—up until local noon. After that the solar energy incident on solar panel **B** goes back down until it reaches zero at dusk.

For this special case solar panel **B** receives approximately 44% as much total sunlight energy over the entire day as solar panel **A**. But remember, that only applies to this special day and location on the Earth. Every other solar panel installation with fixed panels will get less energy than that—and in most cases, a whole lot less.

For sunlight coming directly down through the atmosphere at right angles to the ground, about one-quarter of its energy is absorbed before it reaches the ground. This is what is termed in the industry "one atmosphere's absorption," or AM1.0. When the sunlight travels through the atmosphere at an angle it has to go through more air and thus more of the sunlight gets absorbed. At an angle of around 48° degrees from straight down the distance traveled is about half-again as far and the absorbed fraction rises to about one-third. This is what the industry has settled upon as its standard amount of air absorption of sunlight they will assume when specifying the performance of a commercial solar panel, and they call it AM1.5.

Of course, the actual intensity of sunlight that reaches a solar panel at ground level varies from day to day and hour to hour, depending on where it is located, and it will also be affected by the level of pollution in the air, not to mention weather like rain, snow, or simply clouds that obscure the sun.

Furthermore, if the solar panel is not rotated constantly to keep it facing the incoming sunlight, the amount of sunlight it will intercept will decrease and its electric power output will fall due to the geometric effect. Most ground-level solar installation do not attempt to "track" the sun as it appears to travel across the sky. Some will follow its daily motion but ignore the seasonal change in the apparent path of the sun across the sky. Only the most sophisticated (and expensive) installations do two-angle tracking to keep the solar panels always at right angles to the incoming sunlight.

In addition to the direct rays of the sun, a ground-level panel will also get some additional solar energy from the blue sky around the sun. One estimate is that this increases the solar panel's output, on average, about ten percent.

Solar Photovoltaic Installations Take a Lot of Space

The previous figure showed you one special case, and can be thought of as a view of the solar panels from the north (with east on the left and west on the right). In the next figure I show you another case, looking from the west (with north on the left and south on the right) at a different location, specifically at 34° north latitude.

Here I am showing a couple of ways one might arrange fixed solar panels on the ground for this latitude. The left side has the panels tilted up just enough so that at the summer solstice the sun will come down at right angles to the solar panel. This maximizes the power the panel will generate on that day, which also is when the sunlight is coming through the least amount of atmosphere and thus is at its brightest for the year. (And the weather in most places is more likely to be clear than in the winter time.)

I've shown the sunlight as a number of parallel beams. This is so you can count the number of beams that fall on the solar panel to see how much the geometrical effect reduces the panel's output when the beams don't hit the panel at right angles.

The right side of this figure shows panels that are tilted enough that the sunlight hits the panels at right angles on the equinoxes. This means that the geometric effect will reduce the panel output equally at the summer and winter solstices. But because of the greater tilt in winter, the panel will produce less electricity then that at the summer solstice.

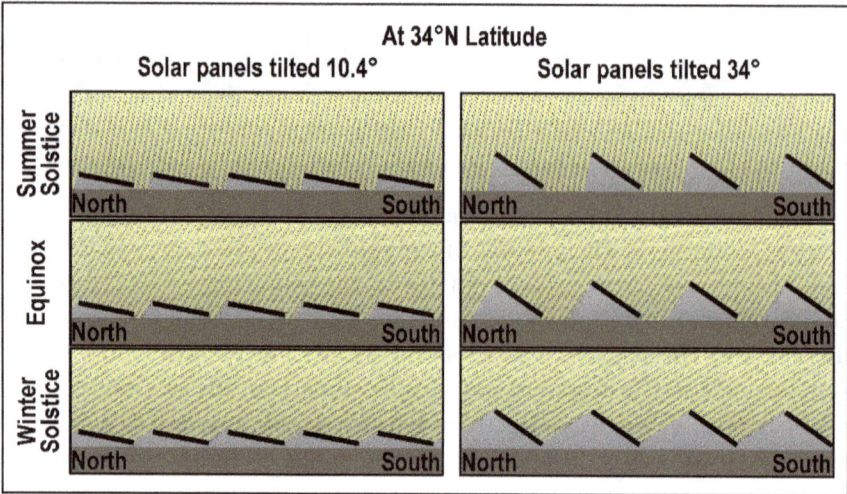

At 34°N Latitude

Solar panels tilted 10.4° — Solar panels tilted 34°

You'll also notice that the solar panels are spaced apart. This is necessary to make sure that they never shade their neighbors. You can think of the way that solar panels combine their output to support an electrical load as being much like the way that the legs of a table share the load of whatever is piled on top of the table. If you have too big a load on the table, you can add another leg. But if that other leg is even a little bit shorter than the rest, it won't take *any* of the load. Similarly, if one of the solar panels is even partially shaded, it may be unable to contribute any electrical power to supporting the system's load.

The left side of this figure shows five solar panels; the right side shows only four. That is because with the higher tilt, the winter solstice shadows reach farther and so fewer panels will fit in the same space.

The higher tilt does produce a more-nearly even output over the year, but at the cost of a lower overall output. This is why most installations use something like the smaller tilt I have shown here.

Not tracking the sun is generally deemed acceptable for ground-level systems since a tracking system costs more, requires more maintenance, and— to keep the adjacent solar panels from shading one another as they are tipped—the panels must be spaced more widely apart. This reduces the total number of panels one can fit on a given roof.

One could, of course, put the panels right next to one another if they were all parallel to the ground. This would allow one to get the most benefit possible from the sunlight falling on a given space, but it would cost a good deal more for all those additional solar panels. [26]

I chose 34° north latitude for this figure since that is where I live, near Los Angeles, California, USA, but also because that is about the latitude of Tokyo as well as of a location midway between Shanghai and Beijing, China; the northern edge of the Libyan dessert; Beirut, Lebanon; and the northern portion of India. Thus it represents the heart of the "sun belt" in the northern hemisphere where there are lots of people living and lots of sunlight, and it represents the region of Africa where many Europeans have said they'd like to place vast solar arrays to serve their needs. In short, it may serve as a fairly typical example of a solar installation at ground level.

Solar panels located outside the atmosphere will be quite different. They will only get the direct rays from the sun (there is no scattered light from the rest of the sky), but those direct rays are all full strength all of the time from dawn to dusk. A very simple arrangement of multiple panels can be built that will easily and quite economically give full two-axis tracking, letting the panels produce lots more electric energy than similar panels at ground level. In my longer book (see footnote on page 50) I'll show one such design and present the relevant calculations to show the ratio of output of these solar panels to what they would produce in a typical static ground-level installation.

An approximate way to summarize the difference is to say that ground-based solar panels, depending on where they are located, and assuming they don't track the sun, are seeing sunlight that, at its brightest (at local noon), is on average perhaps only one-half as strong as outside the atmosphere. Further, these panels will only intercept an average of perhaps two-thirds of the sunlight because of the geometrical effect, and they'll see it for only about one-third to one-quarter of the day. This is why I say that solar panels located outside the atmosphere are roughly ten times better at generating power than those on the ground.

The Real Problem With Ground-Level Solar Power

The real problem with ground-level solar power isn't just the loss of efficiency of the panels from not seeing as bright sunlight in the middle of the

---

[26] You *could* design a system with a single huge panel that held all your solar panels closely spaced side-by-side and then tip and tilt that huge panel so it always faced the sun. But then, for any moderately large installation (10KW or more), that huge panel would be ugly, have a huge wind load (requiring a correspondingly huge support), shade your neighbors much of the day, and as a consequence your city would probably refuse to give you a permit for the installation. So, practically speaking, that isn't feasible for most installations.

day. Nor that they only see bright-enough sunlight to generate significant power for a few hours each day—and then only when the weather is good. The real problem is that we need to have energy at all hours of the day, this means that you won't get to use all of the energy your solar panels do generate for the various tasks for which we use energy. Instead, a good deal of it will be used up making that energy available when and where it is wanted.

You may have read that the managers of electrical distribution grids aren't too concerned by this. That is because at this point there is no likelihood that renewable power (from wind or solar) is likely to become a substantial fraction of the total any time soon. If they tried to run an electric grid today on nothing but renewable power they would be in serious trouble!

And yet, as I hope is now clear to you, it is essential that humanity *completely* stop burning stuff to get energy. If we were to try to meet all of our energy needs for all purposes in any given region of the planet with electricity generated from sunlight, using a ground-level system of solar panels, then we'd have to somehow get that energy to the users at the times they need to use it.

You cannot store electricity for hours or days in the quantities that would be needed. So, apart from those few hours when the local solar panels are working well, one would have to depend on some combination of two strategies:

- Storing energy in a non-electric form, and/or
- Bringing energy in from some distant location where the sun is shining brightly at that time.

One proposed way to store energy is to use the batteries in electric cars. That might be helpful to a small degree, but I doubt we would ever have nearly enough of that sort of energy storage to power all those cars when they are moving around and in addition have enough storage to provide for all the other uses of energy for the many hours of the day when there is little or no sun shining in that region. And if we could do this, it would cost a huge amount. Still worse, batteries only last a few years (typically less than ten), as opposed to the 25 to 30 years that a solar panel will last.

Another way that has been proposed is to pump water uphill when the sun is shining, and then let it run downhill through a turbine to generate electricity in the darker hours. Doing this depends on having a handy source of water and a suitable hill with a large enough reservoirs both below and far enough above the turbine.

But really, any storage mechanism will lose out when compared to solar panels located outside the atmosphere and interconnected in the way I am going to describe below, because the storage and retrieval processes are inherently inefficient. So a considerable amount of the energy one can generate from ground-level solar panels will be lost in those processes. Only a fraction will actually be available for use in the non-brightly-sunlit hours of the day.

The second solution requires an efficient way to carry solar-generated electricity from where it is generated to where it will be used. And this transportation will have to be over some very long distances—up to halfway around the Earth for energy uses that happen at night when the only places where the sun is shining are that far away!

There is one way to do that sort of super-long distance electric energy transport with 100% efficiency: Use a superconducting cable. Some power companies are already doing this, on a small scale. There truly is *no* energy lost in carrying the power along the cable. But the catch is, the cables are only superconducting if they are kept sufficiently cold—at a temperature that must be less than 80 degrees Kelvin (that's negative 193°C, or negative 315°F) for all the presently known superconductors. It takes a lot of refrigeration (which uses energy) to keep the cables that cold. So, while the cable doesn't itself burn up any energy, keeping it cold reduces the overall system efficiency.

The inevitable conclusion is that no ground-based solar system can generate all the energy needed for all the uses in any given region of the Earth without incurring a significant efficiency penalty dealing with the fact that the sun only shines in the daytime—and even then only shines for a few hours brightly enough to be useful.

In contrast, as I shall show you in the section titled "Getting that Electric Energy to the Consumer Efficiently," starting on page 71, solar panels located on top of the trestles can do this job with almost no loss at all, including supplying all the energy we could use anywhere on the globe at any time of the day or night.

So, although the solar panels on top of The Platform may only generate electricity just a few times better than their most-favorably-located and optimally-designed ground-level counterparts, when one considers the losses ground-level generated energy incurs in distribution, the solar panels on top of the Platform combined with the super-efficient distribution system that can be built up there will give that combination an advantage over the ground-level system of at least a factor of ten, and quite possibly a good deal more.

Keeping the Solar Panels Cool

Next, there is the matter of waste heat. Solar panels don't convert all the sunlight they absorb into electricity. Typically they convert anywhere from

10% to 30% at most. (Typical solar panels used today in terrestrial applications convert at best about 20%.) The rest of that energy simply goes into heating the panel.

At ground level that waste heat goes into the environment by a combination of radiation, conduction, and convection. And since the electricity the panel generates usually gets used somewhere nearby it also gets turned into heat in the environment. Thus, solar energy generation systems do nothing to alter how much energy from the sun arrives at the ground level.

Out in space things are *very* different. There is virtually *no* atmosphere to carry away that waste heat. So, up there we will have to provide some sort of cooling to the panels to keep them from overheating (which would at least cause them to generate less electric power).

NASA uses a very simple approach for the solar panels on their spacecraft. They don't worry about cooling the panels. Instead, they let them heat up as much as they must in order to re-radiate all of their waste heat out into space. If you assume both the front and the back of the solar panel are black and exposed to space (with only the front receiving sunlight) and assume an electric power generation efficiency of about 20% then for one square meter of solar panel 1.365kW of solar energy gets absorbed, you get about 273W of usable electricity, and the remaining about 1,092W is waste heat.

The Stephan-Boltzmann law for radiant heat output from a black hot object says that each square meter of hot surface will radiate outward a total of $5.067 \times 10^{-8} T^4$ watts at a temperature of T measured in degrees Kelvin. (That is the unit we use on the absolute temperature scale. On that scale absolute zero—the lowest possible temperature of anything—is $0°K$. To convert a temperature in degrees Kelvin to one in Centigrade just subtract 273).

The reason for specifying a black object is that anything that absorbs radiant energy efficiently will also radiate it efficiently. A white object might bounce off 90% the radiation falling on it, but it also would radiate away only 10% of the amount that a similar black object would emit at the same temperature. For this reason, in order to capture as much sunlight as possible, solar panels are very black at least on the surface that faces the sun. To help keep them relatively cool, NASA makes their solar panels black on the backside, too.

After the solar panel has been sitting in sunlight for even a short while, its temperature will rise until the amount of energy it radiates exactly matches the amount it receives from the sun, less the energy that gets carried away as electricity. Deep space is really cold. (On the absolute temperature scale, which is the relevant one for this purpose, room temperature is about $300°K$ and deep space is about $3°K$.) Almost no radiant energy comes toward us from deep space. So this radiant cooling is really efficient, at least when the solar panel is far from the Earth.

To see just how well this works, you just divide the waste heat from our hypothetical one-square meter solar panel (1.092kW) in half, since half of it leaves via the front and the rest via the back, and plug it into the Stephan-Boltzmann equation and solve for the temperature. One quickly calculates that the equilibrium absolute temperature of the panel will be about 322°K or 49°C. (At this temperature the panel production efficiency only falls about 5% due to the warming of the panel above its rated operating temperature of 25°C.)

This strategy works wonderfully well for the solar panels on space craft sailing far from the Earth. But it won't work quite as well near the Earth, since the Earth is itself a warm body and puts out radiant heat accordingly. So instead of simply radiating away heat out the front and back sides of the panel, it will also be picking up heat from the Earth (which is, on average, at a temperature of about 16°C (289°K).

Our solar panels will be at an altitude of 100km, but that is really quite close to the Earth. If you were up there, looking around, you'd see the Earth filling about half of your field of view. Only the other half would be cold deep space (plus the moon and sun, etc.)—and, in fact, only a small portion of the Earth would be visible. So what matters isn't the average temperature of the Earth as a whole, but rather the average temperature of that portion of the Earth that is visible.

Any object that "sees" nothing but the Earth, and that has no other way to get or give away heat, will eventually come to the same temperature as the portion of the Earth that it "sees." At that point the amount of radiant energy that is flowing away from it to the Earth would equal what is flowing from the Earth to it. We know how much heat flows away from a black object at a specified temperature by radiation: Just use the Stephan-Boltzmann equation and it will tell you. So, we can easily conclude that the amount of energy a black object receives from earthshine is equal to the amount that object would emit if it were at the temperature of the Earth times the fraction of the object's "field of view" that is filled by the Earth.

A square meter of surface of a black material that is at the Earth's average temperature (about 16 degrees Centigrade, or 289K) will emit 353 watts of radiant energy. So a one square meter solar panel that is black on both sides and is about 100km above the surface of the Earth will receive, on average, 176 watts of earthshine in addition to the sunlight that falls on it. If it is above a particularly hot portion of the Earth's surface (think a dark-colored tropical region) where the surface temperature might be as much as 40°C, it will receive 243 watts of radiant energy.

Regrettably, the earthshine doesn't increase the electricity output of the panel. The photons in that light are too red (have too little energy) to cause any electrons to break free from their normal place and move through the panel—which is how visible light or ultraviolet light makes electricity in a solar panel.

Adding either of these numbers for the earthshine absorbed by the panel to the 1,092 watts of waste solar energy, and remembering that the total heat will be emitted equally from both sides of the panel, we can compute that the solar panel 100km above an average portion of the Earth will warm, on average, to about 334K or 61°C. Over the hot region that would rise to 339K or 66°C.

These aren't super hot temperatures; the solar panel will still work, albeit a little (about 8%) less than its rated power. But this is hotter than I'd like, and besides, I have a couple of very important uses in mind for all that "waste" heat, as I'll explain later in this section and in Part Five of this book.

I am proposing that we keep the solar panels cooled below their specified working temperature all the way to 0°C (273K). We can do this by passing a cooling fluid over the backside of the panel and then carrying that heat off to a radiator that is shielded from both sunlight and earthshine. (In a moment I'll tell you how this can be done even when the available radiator isn't quite large enough to keep up with the solar waste heat during the daytime.) Cooling the panels this much should raise their electricity output by about 5%, but I'll stick to the numbers I've used above (which assume 20% efficiency in generating electricity from the received solar radiation) in the following calculations to be very conservative in my cost-benefit estimates.

The solar panels would be built as a sandwich with the active, sunlight absorbing (and heat re-radiating) surface on one side, the cooling fluid layer in the middle, and a very effective heat shield on the back side (to keep earthshine from warming it on that side).

The radiator can be placed behind the solar panel (relative to the sun's location) so that it will always be inside the panel's shadow of the sun. By using a couple of radiators there pointing in opposite directions we can choose to flow the coolant only through whichever of them is most pointed away from the Earth.

Early in the day, when the solar panel is pointed more-or-less due east (to face the sun), one of the radiators behind it will point up toward space and the other will point downward at the Earth. In the afternoon this relationship gets swapped. In the middle of the day each radiator will see some combination of Earth and deep space. And so then there will be some earthshine added to the waste heat that must be dissipated. (An additional white panel on one or both sides of the solar-panel-and-radiators module can shield these radiators from the earthshine coming from either the north or the south.)

You might think that this would make the temperature of the solar panel fluctuate during the day. As long as those fluctuations aren't very large, that might not matter.

Where the fluctuations might be too large (mostly at low latitudes), I propose we do one additional thing that will let us keep the solar panels at exactly 0°C (or just a bit lower) all of the time. This extra step consists of adding to the module consisting of the solar panel and its associated radiators a reservoir of water and ice plus some air at one atmosphere pressure (the air gives room for the water to expand as it freezes and its pressure makes the freezing point stay at 0°C), with the coolant running through this reservoir in addition to or instead of through one or more of the radiators.

During the night, when there is no sunlight falling on the module, the space-facing radiator (which is at 0°C) will radiate away 281 watts per square meter. Running the coolant through this radiator and the water reservoir, but not through the solar panel, will chill that water until it all freezes. Once that happens the reservoir temperature will fall even further, but once it does, you can start flowing the coolant through both radiators and the earthshine from the dark Earth below will keep the module from falling in temperature any further than about 243K (-30°C).

Then, at dawn you switch the coolant flow to go through the back of the solar panel, through the space-facing radiator, and through the ice reservoir. The sunlight falling on the module will quickly warm it up to 0°C at which point the ice will start melting. During the day you circulate the coolant through the solar panel, then through whichever is the most-nearly space-facing radiator and the ice/water reservoir. (If you want to avoid sending any of the waste energy toward the ground during the middle of the day you can just circulate the coolant through the solar panel and the ice reservoir.)

Each of the radiators will have to be made enough larger than the solar panel behind which they are hiding from the sun to let this strategy radiate away all of the waste heat each day in every season of the year. I'll show you a couple of ways in which this can be done in just a moment.

For panels at 50° latitude, either north or south, the length of the day will vary from a low of eight hours to a high of sixteen hours. So by making the radiators large enough that in 24 hours they, plus the solar panel itself, can dump at least as much waste solar heat (plus any earthshine) as arrives in 16 hours, they can keep the module cool every day of the year. At anything other than the summer solstice they'll radiate more heat than comes in, thus cooling the module below 0°C, but with the strategy I just gave you the module temperature can be kept from falling too low. This is the approach I recommend for all the solar panels in the in the tropics or lower-latitude temperate zones.

At high latitudes the day/night length difference becomes more extreme. In fact, in the polar regions there is only one "day" per year, with the sun being above the horizon for half a year and then below the horizon the other half of the year. So in these regions we wouldn't use the ice reservoir strategy at all. Instead we could just size the radiators to keep the solar panel cool enough all day (and here it is easy to shield the module from earthshine, so the waste heat will be constant all day long), and use the earthshine to keep the module from cooling off too much at night.

To keep the module at 0°C simply by radiating away the waste fraction of the incoming sunlight energy we'd need a total radiating area around four times the area of the solar panel—assuming the solar panel converts 20% of the sunlight to electricity. Since the solar panel itself will radiate heat out into space, we need only about three times the solar panel's area of additional radiator. (This is the design we'd use for the near polar region modules.)

To keep the module at or below 0°C with the ice reservoir design, we need only have enough radiator to get rid of the day's waste heat before the next day begins. With a 12 hour day and 12 hour night (near the equator) we'd need an additional radiator that was only slightly larger than the solar panel. For modules to be used at 50°N latitude, with a 16 hour day and 8 hour night, we'd need an additional radiator that had a little less than twice as much area as the solar panel.

How much ice would we need? The "heat of fusion" of water is 80 Joules per gram (one of the largest values for any material). That means that we need 80 Joules (80 watt-seconds) of energy to melt one gram of ice. For a one-square meter solar panel, which produces 273 watts of electricity, we must dispose of 1092 watts of heat all day long. For a sixteen hour day that adds up to about 62.9MJ, for which we'd need about 786 kilograms of ice per square meter of solar panel, or 2.88 kilograms of ice per watt of electricity generated by the solar panel. That is at 50° latitude; closer to the equator we'd need less ice, so this is the worst case. At the equator we'd only need about 2.16kg of ice per watt of electricity generated. (This is a slight exaggeration, since during at least most of the day, as well as all night, the space-facing radiator will be throwing off heat, so the ice supply need not be sufficient to absorb all of the waste heat all day long. But I'll use this larger number just to show that even that wouldn't be an excessive amount of ice.)

This is a lot of water/ice. It means that the ice/water/air reservoir, with appropriate channels for the coolant to flow through it, will extend about a meter behind the solar panel. Further, this reservoir will weigh up to 50 times as much as the solar panel whose temperature it regulates. Clearly for spacecraft launched as they now are by rockets, where each extra gram costs quite a bit to launch, using this approach would be madness. But since the trestles are going to be supporting not only the solar panels, but also a train system and a number of other fairly heavy items—and the train system will allow us to move very heavy things up to the top of the trestles at a very low cost—that additional weight need not be a serious problem.

Another concern you might have would be the total amount of water we'd have to use. I'll deal with that question in a moment, in the section titled "Is This Approach Scalable?" starting on page 72, where I demonstrate that this global solar power system can scale up enough to meet all of our energy needs worldwide.

The next figure shows two versions of the solar panel plus radiators module. The one for low-latitudes also includes the ice/water/air reservoir.

The manifold layer is where the coolant flows to pick up waste heat from the solar panel. The coolant is then pumped (pump not shown) through valves (also not shown) to one or another radiator and/or through the ice reservoir.

The layer marked MLI is a "space blanket" like multi-layer insulation (one of NASA's many innovations) that can very effectively stop any heat from passing through it. This makes sure that all of the heat in the solar panel is carried away by the coolant, and that no earthshine heat can leak in to the solar panel from its backside.

The figure shows the side shield closest to the viewer as transparent. This is only so you can see what it behind it. In practice those side (and for the low-latitude design, rear) panels are very white and well insulated layers to keep earthshine out of the module, and in particular away from the front surfaces of the radiators whenever possible. In the high-latitude design the two radiators are labeled A and B. One of them would be used in the morning hours and the other in the afternoon hours. In the low-latitude design there are also two radiators (not labeled) used in the same manner.

The bent configuration of the two radiators in the high-latitude design shows one way to help keep earthshine off the space-facing radiator for more hours of the day, as well as how to fit more radiator area into a smaller module length.

Tropical and Low-Temperate Zone Solar Module Design

Near-Polar Solar Module Design

Perspective View

Solar Panel
Radiator
Ice/Water/Air Reservoir
MLI Manifold
MLI Manifold
Side View
Top View

Radiator A
Radiator B
Side View
Top View

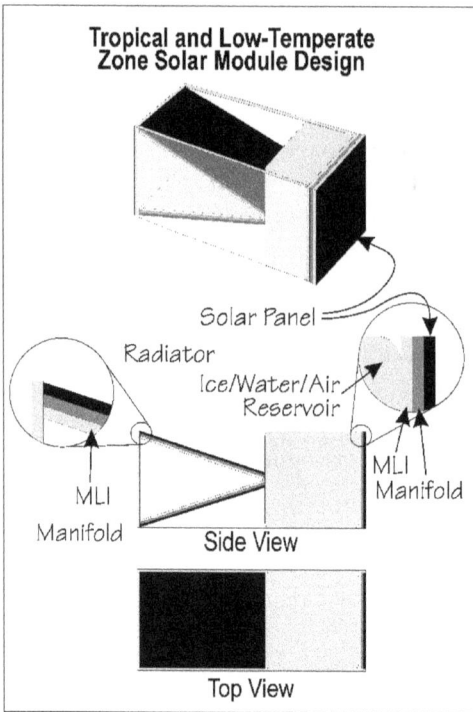

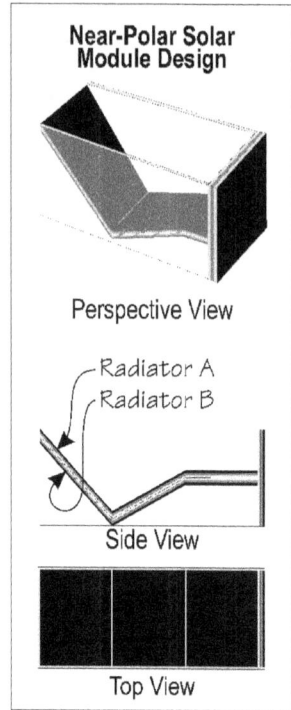

Getting an SRM Benefit From the GEG

The solar panel cooling strategy I just described means we can dump all that waste heat very easily without having to use any substantial energy to refrigerate the panels. We just need enough energy to circulate the cooling fluid.

By dumping all that heat to space we can effectively lower the amount of solar energy that comes into the atmosphere. This is a new way to do something like the Solar Resource Management (SRM) geo-engineering schemes I discussed above, but this time doing that without any substantial penalty in terms of reduced quality of life, nor any requirement to keep paying for it year-after-year. Furthermore, this approach has some additional very useful features that I'll discuss in the section titled "Component #4 "Global Climate & Regional Weather Control" (GCRWCS)," starting on page 95.

Because the solar panels and their associated radiators will be relatively small and very far away, they will be invisible to a person on the ground. Thus they will cast no shadows on the ground, not make the sky look white in daytime (as putting sulphur dioxide up there would) and they won't block any appreciable starlight at night.

That takes care of generating the energy. What about getting it down to ground level where the consumers are? Up to now that has been the big stumbling block for any scheme to use solar power generated outside the Earth's atmosphere.

Getting that Electric Energy to the Consumer Efficiently

You have to understand that electric power distribution systems on Earth now waste a significant fraction (just 6% in the US to a high of 74% in the Republic of the Congo[27]) of the power they generate by burning it up in the wires they use to deliver it.

As I mentioned above, a few utility systems have started to use superconducting cables to get around this loss. Superconductors are materials that can carry electricity with absolutely no loss. But they only do that if they are kept sufficiently cold. So these utility companies have to use some of the energy they generate to run the refrigerators that keep the cables cold enough. For a short cable, this isn't a bad solution; for a very long cable (tens of thousands of kilometers long, as I was suggesting might be needed for a hypothetical fully ground-level solar powered future) the cost to refrigerate the cable could be prohibitively large.

Up in space we, once again, have a big advantage. As is the case at ground level, these superconducting cables will need to be housed inside a long Thermos™-type [vacuum insulated] bottle (or for higher efficiency, inside two nested vacuum-insulated bottles) to minimize the amount of heat that will leak into them from sunlight or earthshine falling on the outside of the cable housing or from conduction through whatever atmosphere there still is at that altitude. So while we must dispose of kilowatts of waste heat from each square meter of solar panels, we may need to dispose of only a rather modest number of watts of heat from each kilometer of superconducting distribution cable.

Once we have the cables shielded in that manner, removing whatever small amount of heat leaks in will be easy. If we use the double-nested thermos bottle approach (quadruple-wall vacuum isolation) shown in the figure we can cool the intermediate layers with liquid nitrogen and cool the

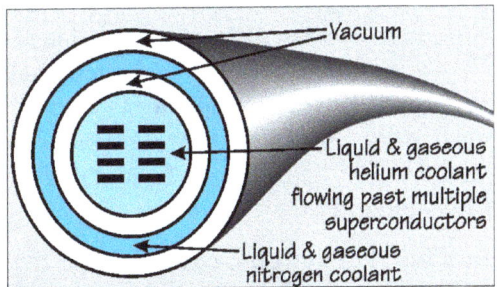

Vacuum

Liquid & gaseous helium coolant flowing past multiple superconductors

Liquid & gaseous nitrogen coolant

---

[27] See: http://www.nationmaster.com/graph/ene_ele_pow_tra_and_dis_los_of_out-power-transmission-distribution-losses-output.

innermost cavity (where the superconductors are) with liquid hydrogen or helium. Both the nitrogen and the hydrogen or helium would be chilled using radiators pointed toward deep space (and, of course, shaded from both the earthshine and sunlight), getting the gas cool enough to condense into a liquid. Then that liquid will either be allowed to flow, or forced (if necessary) through the vacuum-insulated chamber where, as it picks up heat, it will boil off. That boiled off gas will then return to the radiators where it can dump its heat and return to liquid form.

One easy way to make a group of radiators that are shielded from both sunlight and earthshine is simply to lay them out in a line oriented North-South and put an equally wide shade up above them which can be rotated during the day so that its shadow is always covering the radiators. The radiators would have to have a very good insulating layer on the their backside which would be warmed by earthshine. This would probably be a version of the NASA-developed multi-layer-insulation (MLI) materials that they routinely use—and a version of which have now come to consumers as "space blankets."

If we can avoid the need for a pump by using a "heat pipe" design we might even be able to avoid that modest expenditure of energy. This might be particularly easy on the downlink portion of the GEG's interconnection system.

Even if we do need to pump the coolant through the cables, only a trivial amount of energy will need to be expended to circulate the coolant gases through the cable sections and their associated radiators compared to what would be required to refrigerate the coolant down on the ground.

In any case the GEG will be able to deliver virtually 100% of the power it generates from wherever around the earth it does that generation all the way down to the ground near the ultimate consumer of that power.

Also, it might be possible not only to supply electricity to the local utility companies at each place where a cable comes down from the top of The Platform, it may be feasible also to supply additional coolant for extensions of the superconducting cables away from that downlink. That could let those local utility companies operate their systems far more economically than would otherwise be the case—and might make their price for distribution radically lower than it is at present.

Is This Approach Scalable?

You might question my assertion that we could generate enough electric power by this approach. Here are some numbers to help you understand why I make this assertion.

The radius of the Earth is approximately 6.368 million meters. So the cross-sectional area of the Earth (which is the amount of sunlight that it intercepts) is $\pi R^2 \cong 127,000,000,000,000m^2$ (127 million-million square meters, or $1.27x10^{14}m^2$). And each square meter of that area receives 1,365 watts of sunlight all of the time. (There are some slight variations in the sun's output over time, but for my present purpose they are insignificant.)

That means the sun pours a total of about 173,000,000,000,000,000 watts of solar radiation onto the Earth. This is 173 thousand-million-million watts. The proper engineering abbreviation for that is 173PW (peta-watts), or to put it terms of another useful abbreviation, it is 173,000TW. (TW stands for tera-watts, or millions of millions of watts.) Since we are putting our solar panels outside the atmosphere, all of that intense solar radiation which hits the panels is soaked up, with nothing lost by bouncing off the atmosphere or being absorbed within it.

Today the average person uses about 20.3MWhr per year. (That's the average over all 72 countries for which I could get energy use numbers—a group that includes about 92% of all the people living on Earth.) Since there are, on average, 8,766 hours in a year (24x365.25, to account for leap years), this works out to an average energy consumption of about 2.3kW. There were (using the latest numbers in the CIA World Factbook) 7,088 billion total people on the planet early in 2013, meaning that in total they now use about 16.4TW of power (on average). I'm assuming the solar panels only convert 20% of the sunlight they receive to electricity, so to generate 16.4TW of electricity, we'd need to intercept about 82TW of sunlight.

Dividing this number by the total sunlight power that falls on the Earth, we see that we only need to intercept about 0.047% of that sunlight to supply all the power we are using for all purposes now.

If we look ahead to the year 2100, assume the total world population will be around 11 billion people, and further assume that every one of them will by then have a very high-energy usage lifestyle comparable to what we now have in the United States or Canada (the most energy-intense countries on the planet today), they'll each use about 80MWhr of energy per year, which is an average of just over 9kW per person and a total of roughly 100TW of power worldwide.

Since people are spread all around the planet, there aren't likely to be any "peak" hours for this usage; when usage is up in one area it will be down in others, and with this huge number of well-spread out people all living comparably energy-intensive lives, the usage should average out pretty well over time. [28]

To get this much power from 20% efficient solar panels, we'd need to intercept about 0.29% of the total sunlight hitting the Earth—still not a huge fraction. (For safety's sake, we'd probably want to intercept at least 10% more than this, just in case there are some peak usage periods.)

If the solar panels are spaced away from one another even modestly—as they will have to be to keep from shading one another—then their shadows won't reach down to ground level at all. There might be some dimming of the sunlight briefly as the sun appears to pass over the nearest trestle (mainly from the trestle structure itself), but that isn't likely to be enough to bother anyone. The amount of sunlight blocked by the solar panels will likely be an insignificant fraction of that.

I promised I'd also address the issue of how much water we'd need for the ice-cooling technique. Since that will only be necessary for the solar panels that are in the tropical and low-latitude temperate portions of the Earth, even if we need to generate a total of 100TW of power, I'll assume we will get at least a third of that from solar panels in the high-latitude regions. That would mean we'd need to have enough water/ice to cool solar panels producing around 65TW of electricity. In the earlier section I computed that we'd need between 2.16 and 2.88 kilograms of water for each watt of electricity we'd be generating. Using an average figure of 2.53 kilograms per watt, we'd need roughly 165 million-million kilograms of water. That is 1.65×10^{14} kilograms.

I found several web sites[29] that gave estimates for the total amount of water on the planet (mostly in the oceans), the amount tied up in glaciers, and the amount of fresh water in aquifers, rivers, and lakes. The total amount of water is roughly 1,260,000,000,000,000,000,000 liters. A liter of water weighs one kilogram. So this is 1.26×10^{21} kilograms of water. But not all of that is readily available. Only about two percent of that amount is fresh water, but 80% of that is in glaciers and the ice caps at the north and south poles. That

---

[28] We had better help the now-poor but developing countries to complete their transition to fully developed countries by then. Otherwise the world population will continue to rise which could well present us with even worse problems thereafter. Fortunately, building The Platform with its GEG and HST systems will help those nations develop in ways that will almost automatically limit their population growth. For the reasons why I say this, please look at my companion book (see the footnote on page 6).

[29] E.g., http://science.howstuffworks.com/environmental/ earth/geophysics/question157.htm

leaves just 0.36% of the total water in aquifers where it may be accessible by wells, and just 0.036% in rivers and lakes.

Still, given what a huge amount of water there is in total, even this small percentage of the total that is water in lakes and rivers is still a lot of water: there is roughly 4.5×10^{17} kilograms of fresh water in rivers and lakes. What we'd need would only be about 0.03% of the water in rivers and lakes. Clearly we shouldn't worry about using up too much of the Earth's supply. And once we have nearly unlimited and nearly free energy available, we could easily enough desalinate sea water if the fresh water already available seemed too precious to use (or if we need more fresh water for the larger total population).

My plan will mean using a lot of solar panels. If they are similar to the ones now used on roof tops of homes (in the following calculation I'm assuming each panel has an area of about $1.6m^2$), and noting that half of the panels will be on the dark side of the Earth at any given moment, it will take a total of about 75 billion panels to supply all the energy humans now use collectively, and 458 billion panels to supply the maximum amount of energy I just estimated humans might possibly use by 2100. The actual number of solar panels we'll need will likely be somewhere in between these numbers, since even with full development of every country on earth, not everyone on Earth will end up using energy like the average person in the United States or Canada does today. People in Europe now use only about half that amount.

That is one really major solar photovoltaic power installation! What about fitting all those panels up on top of The Platform? Will there be room enough?

I've done the calculations and believe that there will be ample room. I haven't the space here to give the details, but those, and a lot of other details on other aspects of my proposal, can be found in my longer book on this subject (see footnote on page 50).

How Much Will This Solar-Generated Electricity Cost?

Solar panels have been dropping in price rather dramatically, even as their performance and service lifetime keep going up. Recent history suggests the price will fall at least several-fold more as this market matures. If the GEG is built, that will produce a huge demand and will force more standardization and automation in the panel production which will almost certainly lower the cost even more.

Still, I can make a useful, albeit clearly a *very* conservative, estimate of the probable cost of electricity from the GEG by using today's retail pricing of panels and assuming they will perform in space about as they do on the ground—apart from the fact that they'll see a lot more sun more of the time.

For the following calculations I am using figures appropriate to a fairly large solar photovoltaic installation on a residence or small business. Utility scale PV installations are likely to be more efficient, so my cost estimates are surely over-estimates.

Of course, we won't just be putting up solar panels. I propose that we build modules that include the solar panel, the associated radiators, valves, pumps, control electronics, and perhaps ice storage units. These modules could be created in a standard design and in very large numbers, making them rather inexpensive to create. Still, I'll assume that the module price is about double the price for the solar panel itself.

As I was writing this section (in late 2013) the retail price of solar panels was around $1.00USD per nominal watt. (By nominal watts I meant the amount the manufacturer specifies that panel will generate under "industry standard conditions.") Each module would use several solar panels, and each array would use a large number of modules.

The design I'm suggesting would allow the modules to be snapped onto a simple support structure that would hold the modules, rotating them all day to face the sun and tipping them in a North-South direction to follow the sun as the seasons change. The cost for this structure would be at most about 10% of the cost of all the modules it would support.

Similarly the wiring and inverters (if we desire to use the power as AC instead of DC) may add another 20%. So the overall cost for the system would be (at today's prices) about $2.60USD per nominal watt.

I have not included anything for designing the installation, nor anything for labor to install it. These are major components of a terrestrial solar power installation. And for a utility scale installation there also is the matter of buying enough land and getting building permits, etc. Since the GEG is going to be built up on top of The Platform's trestles whose purpose includes supporting the GEG, the space is readily available. The installation will be totally invisible from the ground, so there shouldn't be any complaints "from the neighbors." I expect the design to be standardized so its design cost will be spread over a global scale installation, and thus be insignificant on a per watt basis. The installation and maintenance will likely be done largely by robots and will benefit from enormous economies of scale. This is why I have not allowed anything for these aspects.

A SunPower X-series panel (to cite one of the best models, at the time of this writing, from one of the better United States manufacturers) is guaranteed to produce at least 95% of its rated power output initially, and after 25 years at least 87% of its rated power output. They indicate that the output will decline more-or-less linearly over that lifetime, so the average output over time will be at least 92% of its nominal power output.

Since the GEG panels are going to be outside the atmosphere, they will see full-strength sunlight on average 12 hours per day (averaged over the seasons). And they will generate more energy per hour than their specifications indicate. This is because they will see more light than the industry uses for its specification of solar panel output power. The industry uses a standard test setup that assumes 1kW of incident light per square meter with a 1.5 atmosphere absorption and with the panel held at 25°C. Thus the panel in the standard test setup sees only about 690 watts per square meter, as compared to the 1.36kW per square meter that our solar panels outside the atmosphere will see. This suggests that if the module had enough solar panels to produce one kilowatt of power at ground level under these standard test conditions, it would produce about twice that much outside the atmosphere. Or a bit more since our solar panels will be 25°C cooler.

Thus, for our purposes, the cost per actual generated watt, including all of the panels, mounts, inverters and local wiring, is more like $1.30USD. I said that the solar panels might well have their output decline over time. Let's assume that over their entire lifetime they produce, on average, only about 92% as much energy as their nominal rating would suggest. That raises our cost per generated watt by 8.7% to about $1.41USD.

These panels are guaranteed to perform at least that well for the first 25 years during which time they will be generating power from full-strength sunlight for 12 hours per day (on average) every day. That means for a total of 109,575 hours. (This is 12x365.25x25.) So, for each $1.41USD we'd pay to get a watt of likely generated power, we'll get that much power steadily for at least 109,575 hours. That is 0.109575MWhrs of energy. Another way to look at this is to divide the number of hours into one million, which give you the value 9.13. This means that the fraction of a solar panel that we'd have to buy to generate one MWhr of electricity over its lifetime, is enough panel to generate about 9.13 watts. And that much solar panel would cost roughly 9.13 times $1.41USD, or approximately $12.90USD. That is the "levelized" cost of 1MWhr of solar power from the GEG.

On page 35 there is a chart showing the United States Energy Department estimates of the "levelized" cost of energy generation for each of several technologies assuming the power plants are to be built in the next several years. The Energy Information Agency publishes a document describing how these levelized costs are calculated. For solar power installation they note that the operating and maintenance costs are minor, so virtually all of the cost is the initial capital cost. That means that the levelized cost for this sort of power generation (as opposed to, say, a natural gas fired plant which incurs substantial future costs for fuel, operations, and maintenance) is substantially just that up-front cost to build the system

Here is an new version of that figure, with the GEG cost added. This figure is an update to the one on page 35, and it shows you why I think of the GEG as providing us with essentially "free" electric energy—especially when you realize that the GEG power is actually likely to be less than one-tenth of what this chart shows, as I'll explain in a moment.

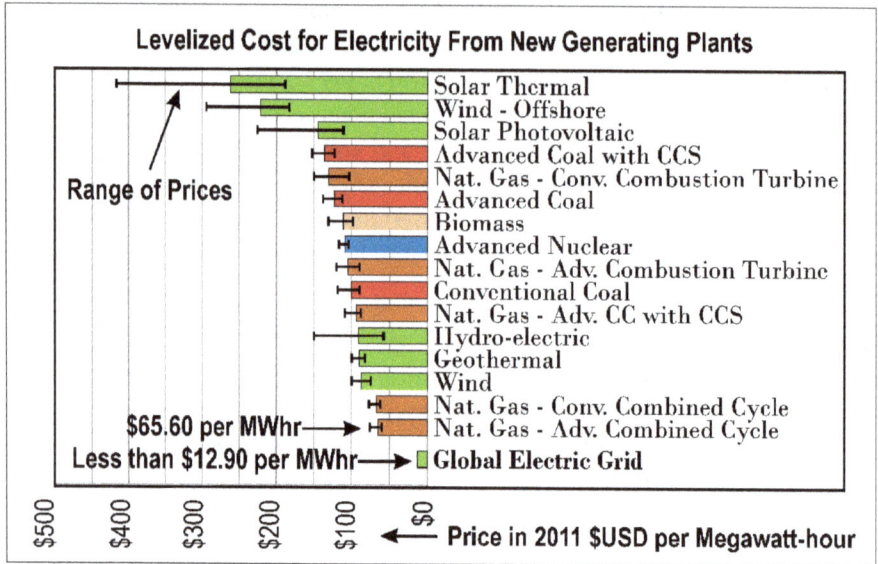

We can use the U.S. Energy Department estimates to see just how much better it would be to generate electric power the GEG way. Simply compare the $12.90USD per GEG-generated MWhr that I just gave you with the U.S. Energy Department's estimate of the average levelized cost to build and operate a utility-scale solar photovoltaic power plant on the ground (using today's technology with a plan for service beginning in 2018) of $144.30 per MWhr. This suggests that the GEG power will be on average about eleven times less expensive than a utility-scale terrestrial solar PV power plant.

Furthermore, comparing the GEG power to the most efficient natural gas power plant's levelized cost of production (estimated by the US DOE at $65.60 per MWhr) and you see that the GEG power will be at least five times cheaper than even the lowest price source we have now (and the GEG would make no pollution, unlike this sort of natural gas plant).

Finally, remember that natural gas power plants are a fully mature technology, with only marginal improvements expected in the next few decades, but solar photovoltaic power is a rapidly developing technology, with prices falling in half roughly every two years. So by the time even the first portions of the GEG are built (at least a decade from now—hence when solar panels will likely cost no more than around one twentieth of what they do now) we can reasonably expect the cost of its electric power to be well more than ten times, and very possibly more than one hundred times cheaper than any fossil-fuel-burning terrestrial power plant.[30] For rough estimating purposes elsewhere in this book I have simply said that GEG power is likely to be at least ten times cheaper than any present source. That is, as you can now see, a very conservative estimate.

Most likely the highest cost to consumers in many countries will be the transmission cost from wherever the GEG delivers that energy to the final consumer some relatively short distance away (either the GEG downlink site, or the end of a superconducting cable run from that site to a closer neighborhood re-distribution site).

Converting All Energy Uses to Electricity

I realize that our societies now are organized around the assumption that most of the energy they use will come from burning fossil fuels. We know how to replace almost all of those uses with electricity, but we don't do that now simply because it would cost us more.

For example, burning fossil fuels (or anything else) to generate electricity and then using that electricity to heat our homes today costs more than heating them by simply burning something (wood, natural gas, or even oil or coal). Electric heat is often more convenient, if cost isn't an issue, but it always less efficient and thus more costly since it takes two steps and much of the energy is lost in the first step. But if we could have that electricity at a far lower cost, then it would make sense to convert *all* of our present and future uses of energy from burning stuff to using only that very inexpensive electric energy.

You might think that this would be hard to do for uses that now seem to demand the use of liquid fuels, such as for airplanes and ships at sea. I'll explain in the next section why, once we've built the new infrastructure I am

---

[30] Of course, this reduction in cost for solar panels will also benefit anyone putting in a terrestrial solar photovoltaic system as well. Still, the cost differential between the ground-based solar PV systems and the GEG will remain at least ten to one, just on efficiency of power generation. Economies of scale are likely to make that ratio even larger. A side note is that if, as I assert, the natural gas electric power stations are using a mature technology that won't be able to make such rapid strides in lowering costs, then solar PV on the ground will soon become the least expensive way to generate electricity apart from the GEG. Still, it couldn't compete with the GEG!

proposing, we'll be able to replace airplanes and ships almost entirely for transporting both people and cargo with a different means of long distance transport that uses only this very inexpensive electricity.

We already know one way to use electricity for trucks and automobiles, with battery powered vehicles. But so far, battery technology hasn't been able to give us the range per "fill up" that a liquid fuel in a conventional internal combustion engine can deliver without requiring a very heavy, large, and costly battery pack with limited life.

The solution to this quandary is already on the market. Wireless charging systems for cell phones are widely available, and similar systems for fully electric automobiles are known to be feasible and will likely hit the market in the next few years.

At a recent Wireless Power Conference I heard a presentation suggesting that each power-using device (a vehicle in this case) would communicate with the power provider, telling it what voltage and current it needed. The provider would then provide just that sort of electric power to that device. And, of course, since the provider then knows whose that device is, it can bill the owner for that energy.

And if we install that sort of wireless charging system on all our major highways and at all our places of business and residences, we could use far smaller batteries in our vehicles. This would make converting to pure electric vehicles superior to using liquid fuels of any kind (fossil fuel or hydrogen) in an internal combustion engine. For one current example of this wireless power approach, see the article referenced in the footnote below about a South Korean bus company's experiment.[31]

---

[31] "World's first road-powered electric vehicle network switches on in South Korea," at http://www.extremetech.com/extreme/163171-worlds-first-road-powered-electric-vehicle-network-switches-on-in-south-korea describes one example of this sort of road-powered electric vehicle now in daily use.

Part Four
Transforming Transportation

Component #3 "The Hyper-Speed Transport System" (HST)

This is my favorite of the several infrastructures I see being built on top of The Platform. One reason is that this was the one I first envisioned, and it led me to the others, and then to seeing just how important building all of them together could be.

I said previously that the transportation infrastructure component is the one that needs the most in the way of special features from The Platform. I'll explain in this section just what those special requirements are, and give at least the basics on how The Platform can be built to provide them.

The Essential Concept

The simplest description of this component is that it is a system of magnetic levitation trains running on top of The Platform. Since at that altitude we can easily cool superconductors, the train will be built using only superconducting coils instead of permanent magnets. This will allow the magnets to be very much stronger than even the best permanent magnets—a feature which will be very useful in places where the track isn't level, which includes places where it will be completely vertical near the track ends at each terminal. (Cooling the superconductors in the tracks near the Earth's surface can be accomplished with very little cost by cooling a fluid up on top of The Platform, and then pumping that cold fluid down to wherever the cooling is needed—the same approach used in the GEG to keep the downlink power distribution cables cold.)

If we also make sure there are no electrically conductive surfaces in or near the track, we can completely avoid any resistive power loss. The only significant source of power loss will be air friction, since even at 100km altitude there is *some* air.

And not only will the train tracks go along the top of The Platform, at each metropolis that The Platform's trestle passes, downlink tracks will curve smoothly down to the ground, and other uplink tracks will curve up from the ground. Each of these uplink or downlink tracks will be enclosed in an evacuated tube—not evacuated to the same low air pressure as up on top of The Platform, but enough to reduce the air friction for even fairly rapidly moving train cars to a level of insignificance.[32]

I have worked out the details of such a train system with a design criterion that a passenger on the train should never feel any acceleration greater than one-tenth of gravity at the Earth's surface. That means that they would feel at most about 10% heavier than normal while they were riding in this sort of train.

Both cargo and passengers will travel in containers that will more-or-less resemble the intermodal containers commonly used today on trucks, trains, and transoceanic ships to carry all manner of goods from place to place on our planet. However, these new containers will have to be built in a much more sturdy fashion, especially those that will be used to transport people. This is because during each trip on this new transport system those containers will rise up out of the Earth's atmosphere and travel briefly in what is essentially outer space. So the containers must be built in a manner resembling today's airplanes and space ships for people transport—air-tight and with suitable life support.

The major difference between this transport system and trains we have to day will be the speed. Right now people traveling across a continent (for example, Los Angeles to New York) expect to spend several hours in the air.

---

[32] Some readers may think that what I am describing is not much different from what has been called the Hyperloop (proposed by Elon Musk) or the ET3 project (Evacuated Tube Transport Technologies). The Hyperloop is a relatively short-range pneumatic tube system, which is very different from the HST system, and offers only a tiny fraction of what the HST system would offer. The ET3 concept does resemble the HST in many ways, but by placing it at or near ground level it doesn't have many of the HST's benefits, and indeed, cannot be made to work as well. In my longer book (see footnote on page 50) I describe each of these alternative proposals and explain in detail why neither is as useful as the HST. Here I just want to point out that they only serve to move people or cargo from one point on the earth to another. They do nothing to address our energy needs, or global climate change.

Going halfway around the world can take a day or more. Containers of cargo traveling on ships across the ocean typically take weeks.

Because of the way the train cars and the containers they carry are built, and because of the subtleties of how gravity and centrifugal force interact, it will be possible to travel between any two places on earth that are connected by one of these new transport links in *under an hour*.

If you have heard about something called the Gravity Train[33], you may think this is similar. It is not. That also would be a great way to get people and cargo from place to place on the earth—if it could be built. But it faces far more severe challenges than those that will face the builders of this new transport system, and further, the Gravity Train does not (and can not) include the solar power generation and distribution system or any of the other infrastructures that are included in my proposal.

Because of the very high speed of this transport system, I propose we call this the **Hyper-Speed Transport** system, or HST, for short.

The Societal Impact of the HST System

The impact that building the HST System will have on all the societies that are interconnected by it will likely be enormous. I cannot go into them in any detail here. But one point I want to make may suffice to hint at the impact in other realms. I'll give you more details on this in my longer book (see footnote on page 50).

I will define a metropolis, functionally, as a region within which there are a lot of people and businesses and where one can get from any part to any other part of that metropolis within a couple of hours. By that definition, once the HST system is in place, **all of the different metropolises it connects will be functionally the same metropolis.**

Think about that for awhile. Anyone living anywhere in a major urban center on the Earth could just as easily work in any other such urban center as they could in the one they live in, and they could also do their shopping

---

[33] Prof. Michio Kaku of CCNY briefly describes the Gravity Train in a video from the Discovery Channel that can be viewed at this URL:
https://www.youtube.com/watch_popup?v=t1gTzc7-IbQ&feature=player_embedded
In my longer book (see footnote on page 50) I will detail why I think we'll never be able to solve the engineering challenges of the Gravity Train, as well as explaining why some of the other far-out transportation proposals now being talked about are also unlikely ever to be built.

and/or get their entertainment in yet other urban centers. The mind boggles at what all the consequences of that might be.

I must hasten to add that we would have some very tough problems to face about customs and immigration controls, human rights issues, employment standards, even national sovereignty before there could be such an easy flow of people worldwide. I'm not a sociologist nor a politician so I'll have to leave devising and implementing the solutions to those problems to those other experts. Whoever does those things, will I hope conduct a robust conversation with all the people who will be affected, which is all of us, before the specific solutions are chosen.

Will It Attract Enough Riders – Can It Scale Large Enough?

These are two obvious questions. And I've considered each of them quite thoroughly. You'll find all of my thoughts on these important topics in my longer book (see footnote on page 50). Here I'll just assure you that (a) the HST system will offer transportation that is so superior, in both its high speed and its low cost, that it will quickly attract almost all of the transport demand wherever it is built, and (b) it will be easy to supply as much transport as people want. I did the calculations for Los Angeles, for one example, and just a few uplink and downlink tracks would suffice to handle all of the people and cargo that now travel in or out of that metropolis. And, in any case, one would need quite a few uplink and downlink tracks to service all the different directions away from Los Angeles that this traffic would want to go.

Some Details of the HST Transport System's Operation

I'll not give you more than just a sketch of the HST operation here. I assure you that I've created a detailed model to make sure that it will work as I am describing, taking into account all of the physical forces that will come into play. For the details you'll have to consult my longer book on this subject when it comes out (see footnote on page 50).

Fooling the Passengers

The great speed with which the HST system can carry cargo or passengers over great distances means that the HST cars must at some times speed up rather quickly, and then later on also slow down equally quickly. The formal physicist language for this is to say these cars must at times have a substantial acceleration.

Acceleration is any change in an object's velocity. The unit for velocity is meters per second. Speed is just the size of the velocity number; velocity itself is a concept that includes that size and a direction. Acceleration is how much the velocity (including direction) changes per second. The unit for acceleration is velocity change per second, so the unite of acceleration is meters per second per second. Some accelerations involve changes in speed; others only involve a change in direction; still others include both kinds of change.

In order not to let the passengers feel the large speed changes of the HST car as anything out of the ordinary, the container they are riding in will hang from the HST car. This will allow the container to swing like a pendulum—even allowing it to swing upside down. During a long trip it will swing in just this manner, but the passengers will never feel it.

As the car rises up the initially vertical track, it will speed up along the track accelerating at a rate of one-tenth of gravity (accelerating at roughly one meter per second per second, or about 3.2ft/sec/sec). This forward acceleration pushes the passengers back along the track and makes them feel about 10% heavier than usual. As the track continues to rise and begins to curve over toward the horizontal, there will be a gradual buildup of centrifugal force. That force pushes the passengers away from the center of curvature.[34] Initially that is mostly to one side of the track (the side away from that center of curvature), but once the track has curved over enough so that it is no longer going simply straight up, the centrifugal force will push only partly to the side and partly upward, thereby partially cancelling out the pull of gravity down toward the center of the earth. Finally when the HST car is all the way up on top of The Platform, the fact that the top is at a constant elevation above mean sea level means that the track will curve around the Earth, and the centrifugal force from that motion exactly points in the opposite direction to the pull of gravity.

At all times we want the passengers to feel exactly as if the only force on them is gravity, but with that force being just ten percent more than they are used to feeling. To accomplish this, the car must be able to swing out from the vertical just enough to keep the floor of the car at right angles to the sum of three forces: (1) Gravity pulling down toward the center of the Earth; (2) centrifugal force pushing away from the center of curvature—and eventually pushing straight up, thereby effectively weakening gravity; and (3) the reaction to the forward acceleration of the car, which pushes the passengers back along the track.

---

[34] Some people think of centrifugal force as somehow different from gravity; they call it a "fictitious force," and say it isn't "real." It arises from a different source than gravity, but it certainly feels just as real. When a person is undergoing accelerated motion they aren't in what physicists call "an inertial reference frame" and any calculations about what they feel and what happens to them must include both the "real" forces, like gravity and the force exerted by the motor on the train plus any "fictitious" forces, like centrifugal force.

This means that as the sum of the centrifugal force and gravity reduces the passengers' effective weight, the HST car must accelerate forward along the track more rapidly until the sum of all three forces is exactly 110% of gravity. (Note: This sum is what we call a "vector sum" so one cannot simply add the magnitudes of the three forces; instead one must take into account the directions of those forces, as well as their strength.)

What I have just described is just like the feeling of riding in an airplane that banks as it goes around a turn. If the bank is at just the right angle, you don't feel as the airplane is tilted at all—it feels "level" even though a glance out the window shows that it is not.

Here is a diagram showing how an HST car carrying a container of people (or cargo) might look at several key points along the path from Los Angeles, California toward Paris, France. This figure only shows approximately the first one-third of the that trip, with the inset showing some of the details of the first 200km of the trip, where the HST track is curving up toward, and finally joins the tracks up on top of The Platform.

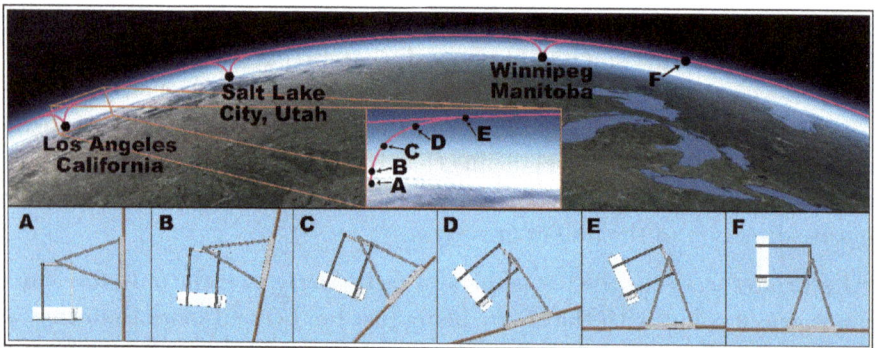

┌───┐
│ **A Suggestion for the Non-Technical Reader:** In the next several
paragraphs I'll explain in detail just what the HST car and its container are
doing at each of the six labeled points (A through F). If you aren't interested
in seeing all of those details, just jump to the next section, "One Trestle Serving
Multiple Cities," on page 88.   │
└───┘

At the beginning of the uplink track (point **A**), the track goes straight up. In the lower six drawings the HST car is the grey rectangle parallel to the track (initially vertical) and with its A-frame (initially pointing out to the left). The container is the white rectangle hanging from that A-frame support.

As the HST car rises and the track starts to curve over—and the speed of the car rises, too—the container will start to hang out to the left a little bit in order to balance the force of gravity (down), the centrifugal force (off toward the left and a bit upward) and reaction to the car's forward acceleration (back along the track).

Point **B** is where the HST car has risen to an altitude of 20km. That takes just 3 minutes and 22 seconds after the car's launch (the start of its rising up the track), and by this time it is moving upward at 204m/sec (456mph).

By the time the altitude is 40km (point **C**) the track has tipped over 45° and the HST car has moved not only 40km up but also 24km horizontally. The container for people or cargo is now hanging out to the left about 21° from vertical. It arrives at this location just under six minutes after launch and is now moving at 471m/sec (1,053mph). This is faster than any train on Earth today, but far from the top speed the car will achieve during its trip to Paris, France, let alone the top speed any HST car could achieve on a sufficiently long trip. (I'll describe the absolute speed limit of the HST system shortly.)

At point **D** the car passes through altitude 90km, moving 926m/sec (2,071mph) a little over seven minutes after launch.

Point **E** marks where the uplink track arrives at the top of The Platform (altitude 100km). The car gets there about 8.5 minutes after launch and it now is moving at 1.49km/sec (3,335mph) and has already moved 96km toward its destination. (Because of the curve of the uplink track, it has risen 100km and traveled sideways 96km while only traveling 128km along the track.)

At point **F** the car has traveled about 3,113km and its speed is 7,843km/sec (17,544mph) which is exactly what is termed "low-earth orbital speed." This means that by traveling along the top of The Platform (which is at a constant altitude and thus is curving in a circle who radius is the Earth's radius plus 100km) and going at this speed one creates a centrifugal force (upward) that exactly cancels the gravitational pull of the earth (downward). Since these two forces exactly cancel one another, the HST car must accelerate along the track with an acceleration equal to 1.1 times gravity (as opposed the 0.1 times gravity acceleration it had at the start) in order to keep the passengers' perceived weight constant. And it also means that the container will be hanging straight out to the left.

Still, to the passengers in the car, it will always feel as if the car is "right side up" and they are simply seated as if in a normal train car. They will, perhaps, notice that their weight went up gradually (over many seconds) at the start of the trip—but most likely by this time they will have forgotten that, as their apparent weight will have stayed constant at 110% of normal for the duration of the trip (up until the last several seconds). They won't feel the tilt of the car or the enormous speed they are traveling.

One Trestle Serving Multiple Cities

You may have noticed the curved magenta lines coming down from the top magenta line to Salt Lake City and back up again to the right, and a similar pair of curved lines at Winnipeg. These show how the same trestle that serves HST cars going from Los Angeles to Paris can also serve cars going from Los Angeles to Salt Lake City or Winnipeg, or others going from Salt Lake City or Winnipeg toward Paris. All of the six cities whose locations are indicated in the figure on page 42 are under or very near the path of the Los Angeles to Paris trestle, and so all of them can be served in this manner by that one trestle segment. (The trestles should follow "great circle" routes from beginning to end. If they do not, there will be large sideways forces on the train, keeping it from accelerating forward as much as it otherwise could.) Also you can see that at Los Angeles there is uplink track for cars going south as well, and the main track on top of The Platform is also extended south, perhaps toward New Zealand.

An Absolute Speed Limit

You might be surprised to learn that, fast as it is, the HST system has an absolute upper speed limit. It is a consequence of my design criterion that the passengers should never feel more than 110% of their normal, sea level weight.

I mentioned previously that going around the Earth at such a high speed can make the centrifugal force greater than gravity. As long as it isn't more than about twice gravity, the sum of the centrifugal force (outward) and gravity (inward) is some amount less than 110% of g [g is the symbol for the force of gravity at sea level]. And so long as that is true, there will have to be some forward acceleration to make the sum of all three forces exactly equal 110% of g. But once the centrifugal force plus gravity equals 110% of g, there can be no additional forward acceleration. The HST train will then have reached its absolute maximum speed.

You might think that would happen when the centrifugal force alone equaled 210% of g. That would be true, except that the HST track is 100km above sea level. The Earth's gravitational pull falls off with the inverse square of the distance from the center of the Earth, so at that altitude the force of gravity is only 0.966g, and thus the maximum permissible centrifugal force is just under 207% of g.

Specifically, after the HST car has traveled on a trip whose total distance is something more than 16,220km for just over 27 minutes it will have gone 8,110km (5,039 miles) along the track, and it will be moving at 11.456km/sec (25,627mph). This is its maximum speed. The ground distance traveled at this point is 8,057km (5,007 miles). The difference between the distance traveled and the ground distance traveled is because the HST car had to travel up to the top of The Platform as well as along the path to its present location. And at the other end it will again have to go about 53km more than the ground travel as it comes down the downlink.

The HST car's speed will stay constant from that point until the car reaches an equal distance past the midpoint of its trip. From then on it will gradually decelerate until it reaches the destination at zero speed. (The container will tilt out in front of the HST car during this phase of the trip, and at the end it will have turned around one complete circle from where it was at the start. Still, the passengers won't feel a thing to indicate that this has happened.[35])

The Necessary Special Features of The Platform

The top speed I have indicated for the HST trains is really, really fast. No train on Earth has ever traveled anything like that fast. The fastest commercial train today is the maglev train that connects Shanghai to its airport which runs at speeds up to 400 kilometers per hour (about 250 miles per hour). Japan, France, and Germany also have some very fast maglev trains. The world record speed for a train is 575 kilometers per hour (which is about 357 miles per hour) set by a French train in 2007.

By comparison, the HST cars will go at speeds up to 11.46 kilometers *per second*. That is equivalent to about 25,500 miles per hour. That is just a tad faster than the escape velocity from the Earth. So if an HST car at that speed were to break free from the track it would fly away from the Earth, never to return.

Right away you can see one essential feature of the HST tracks and cars: They must be built so that the car cannot ever fly off the track. (It may be useful to let the HST car release its container in some situations—as that would be a very nice way to launch a space vehicle. I'll talk more about that in a moment.)

But a more subtle requirement is at least as important. The HST track up on top of The Platform must be really, really "straight." I put straight in quotes because it must be curved just the right amount to curve around the Earth so it can stay at a constant altitude. And it may need to curve very, very gently to the side if the HST car is to move from one track to another track running parallel to the first track, or if the journey requires the car to move from a trestle that follows one great circle route between say, City A

---

[35] For any trip of less than 16,220km the container must somehow shift from hanging back while the HST car is speeding up to hanging forward while it slows down. The details of how it can get from the one to the other position without the passengers realizing it is covered in my longer book (see footnote on page 50).

and City B to one that goes from City B to the City C along another great circle route that is slightly angled from the first one. Even at the high speeds at which the HST cars travel, it would be possible to curve the track to the side as long as the radius of curvature was long enough. (Going around that curvature at these high speeds adds another centrifugal force component to the side, and as a result, the forward acceleration has to be reduced somewhat during that transit of the sideways-curved portion of the HST track. This means that no such portion could be navigated at the HST car's top speed without violating my design constraint.)

How can this be done? The Platform must be built to ensure a very high degree of track "straightness" at all times. (This requirement is relaxed on the uplink and downlink portions of the track since the HST cars are traveling much more slowly there.)

And that means that The Platform must not simply be a static structure. Instead it must be dynamic, able to readjust its shape within seconds whenever it detects that something has moved any portion of the HST track even a little bit out of its proper position.

When might this sort of adjustment be necessary? Any and every time even a little bit of misalignment occurs. And there are many, many things that could cause that sort of misalignment.

One is simply the force of the HST car as it passes along the track. Since the car may be heavy, it may tend to flex the track down (if it is going slowly) or even pull it up (if the HST car is going very fast). And if the track curves to the side, the HST car will tend to pull or push it out to the side as it goes around the curve.

This sort of problem is readily predicted and the correction can be applied proactively as the HST car moves along the track, stiffening the support at each place in just the right way to keep any significant misalignment from happening as it goes by that place.

Another cause of misalignment might be something that moves one of the supports of The Platform down at ground level. A passing truck might vibrate that support. An earthquake could quickly displace it by as much as a few meters in just a few seconds.

Or an asteroid might crash into a portion of The Platform and that would likely wrench it out of shape. (Also a terrorist bombing a portion of The Platform would likely throw at least a portion of it out of proper alignment.)

Fortunately, in the case of such distant causes for misalignment, the very size and height of The Platform will work in our favor. Consider the case of an impact (or earthquake displacement) of some point on a segment of The Platform near the ground.

This disturbance of its shape at the impact point will not affect a distant portion of The Platform at least until there has been time enough for a sound wave to travel through the structure from the impact point to where the HST cars are traveling. Given The Platform's size, that time will be more than a minute even for a location along the HST track directly above the impact point. Furthermore, the adjustments to The Platform's shape will be very minor when expressed as a percentage of its overall height.

Sensors can be distributed all over The Platform to detect problems and communicate their findings to a nearby control computer in a tiny fraction of a second. That computer can command the dynamic features of that and other adjacent sections of The Platform to adjust their shape to compensate for this disturbance. In effect, The Platform will generate a sound wave going down that will exactly cancel out the disturbance coming up toward the track. The energy of the disturbance will simply be reflected back down toward the ground, and the portion of The Platform near the HST track won't "feel" a thing.

Similarly any slow movement of the supports (for example, from tectonic plate drift) can be compensated for by the dynamic nature of The Platform.

I've given this a lot of thought and have worked out many of the details of just how this can be done. Again, I haven't room here to give you those details, but they can be found in my longer book (see footnote on page 50).

A Marvelous and Important Bonus

I mentioned that if an HST car were to release its load when it is traveling really fast up on top of The Platform, that load would be launched into orbit. If the HST car's speed were only slightly above low-earth orbit speed, the load would go into an elliptical orbit around the Earth. And either with some small chemical rockets, or for an even "greener" version, by pulling power out of the Earth's electric and magnetic fields, that load (now a satellite) could adjust its orbit to whatever might be required of it.

This would be a way to replace all of the very expensive and often polluting chemical rockets now used to launch space craft. I've worked out the cost and it is remarkably low. To accelerate a one kilogram load from zero speed to the HST system's absolute maximum speed requires about 18kW-hr of energy. At a cost of $12.90USD per MW-hr that works out to about $0.23USD per kilogram. That is for accelerating that mass to a speed that is actually just over the escape velocity from Earth's gravity. To get just up to low-earth orbit would only cost about $0.11 USD per kilogram. (And remember that my electric power costs were *very* conservative—the actual cost for GEG-powered HST space launches will almost certainly be a small fraction of these values.)

Right now, using chemical rockets, the dream (almost, but not quite yet accomplished) is to be able to launch a payload to low-earth orbit for as little as $1,000 USD per pound (thus $2,200 USD per kilogram). So you can see that even the best dreams of the chemical rocket aficionados today is to get their cost down to something that is about 20,000 times more than the maximum cost I've estimated for using the HST system and the GEG to generate the electricity for the HST train.

The biggest reason for this economy is clear: Chemical rockets are very heavy when they leave the launch pad. Most of their weight is the fuel that they are going to burn up on the way to space (and, for a round trip, they must also carry all the fuel to be burned up upon reentry). So you have to pay to launch anywhere from ten (for a one-way trip) to one hundred times (for a round trip) as much mass as your actual payload.

With the HST car functioning as a "train to space" you only have to pay to accelerate the payload—you don't have to accelerate any fuel. And since the energy needed to accelerate the HST car that carries the space vehicle will be recaptured as that car is slowed down after it has launched the space vehicle, you really only need to pay for the space vehicle's acceleration. (Furthermore, if that space vehicle is carrying people, it almost always will sometime later return those people to earth—and if that is done via the HST system then almost all of the kinetic energy it received as it was accelerated will be recovered as electricity as it is decelerated, thus lowering the price of round-trip space travel even more.)

Of course, once the load or satellite is launched from the HST car, it will be in "free fall." If that load includes people, it might be nicer to have some way to continue to give them that artificial gravity feeling that they had on the way up to the launch point. I've worked out how to do that as well, plus a means for returning satellites (with or without artificial gravity) via the HST transport system. Again, the details are in my longer book (see footnote on page 50).

The true importance of this space launch capability may not be obvious. Clearly, it will be of economic benefit to those who now launch space craft using rockets. And it will be enjoyable for space tourists. But I assert that it will likely be even more important to the rest of us over time.

Every astronaut has spoken of the moving, even spiritual experience they have had when they first went into space. They get the chance to see the Earth from outside. And that has moved many of them to become quite active in helping teach others why we should work to preserve what we have. (You can see some of their quotes in the section titled "A Matter of Perspective" located right after the Table of Contents at the front of this book.)

I think offering this experience to virtually anyone at what turns out to be an extremely modest cost could transform people's attitudes in a very wonderful way. The cost to take a one-way trip to space this way would be well less than $100 USD per person—with most of that being returned to the passenger when they return to the Earth. Furthermore, there would be no need to be especially fit—which, in addition to the cost, is what now prevents most folks from even thinking of going into space.

But apart from these sorts of consideration, there is one more way this capability could be used—and that is to meet a real, critical need we have, and for which no other means of meeting it has yet been perfected.

Meeting a Critical and Now Unmet Need

I'm referring to cleaning up the mess we have created in low-earth orbit. A recent article in Photonics Spectra magazine[36] includes this comment by Creon Levit, chief scientist for programs at NASA Ames Research Center in California:

"If not mitigated, the problem of collisions with orbital debris in space will continue to damage and destroy satellites—and other space missions, like the ISS [the International Space Station]—in Earth orbit. With each collision producing thousands of new pieces of space debris, the result over the long term (around 100 years) could make space operations and space travel impossible."

This drawing is NASA's way of illustrating the problem. The recent movie, "Gravity," is Hollywood's way of dramatizing the problem.

NASA has already set up a program called Light Force to see if they can perhaps push bits of space debris out of orbit by shining light on them from the ground. That might work, but it seems like a difficult and chancy approach that is unlikely to get all the junk out of orbit any time soon. And as those bits fall to Earth some may burn up, but some

---

[36] Photonics Spectra, September 2013, page 44.

may come crashing down and do unknown damage to whatever is in the way.

The HST system offers a relatively easy and inexpensive way to solve this problem. Just launch a container with its front doors open into an orbit that intersects the orbit of some target bits of space junk. Once a container surrounds the junk, snap its door closed until the junk is corralled inside and secured. Then open the door and adjust the container's orbit to go pick up the next bit of space junk. Essentially this would be a space garbage truck picking up the trash as it goes along. And once it is full, it can return to the HST system where it will be caught and carried down to earth to be emptied and then sent up again.

This also addresses one of the concerns I was asked about when I was discussing the GEG. The questioner asked me about the possibility that the solar panels would be damaged by flying space junk. That is a very real potential problem, and it is one that might best be dealt with by using this space garbage collection function of the HST system to clear out all the bits that might otherwise cause havoc for the GEG and any other important things that are up at altitude outside the atmosphere's protection but still fairly near the Earth.

Part Five
Putting a Thousand Thermostats on the Planet

Up to this point you have learned how building the trestles and placing solar panels (and the associated radiators, superconducting interconnections, etc.) up there, plus building a global long-distance transport system there will give us all the energy we need for all uses and do a whole lot more. What it won't do is save us from extinction.

This is because we already have dumped so much carbon dioxide into the atmosphere (and will have done quite a bit more by the time the new infrastructures are complete). Further, that carbon dioxide will linger in the atmosphere for many centuries. So, we can anticipate that the planet will warm overall so much that we likely will become victims of the Earth's sixth mass extinction event.

But there is a way we can compensate for all that excess CO_2 in the atmosphere, plus this new way also will allow us to dial up or down that compensation as necessary, and even allow us to modify regional weather patterns as well. We'll have to learn how to use this new capability, but once we do we may be able to damp down some of the most destructive "extreme weather"—whose incidence now seems to be rising inexorably.

Here's how we can do all that using just what we will have already built (with only a few very minor modifications and additions).

Component #4 "Global Climate & Regional Weather Control" (GCRWCS)

This is the last of the essential components of my proposed new infrastructure. And it is actually just an elaboration of something I already mentioned in my discussion of the GEG.

Do you remember when I pointed out that with a solar panel on the ground all of the sunlight energy that it absorbs ends up being dissipated in the environment somewhere nearby? Some of it goes into the surroundings as heat right away. The rest is converted to electricity and sent to some, usually nearby, motor or other device where it also gets converted to heat.

But with the GEG, only the smaller portion that is converted to electricity will be sent down to the ground. The rest will be radiated away, typically out into deep space.

I pointed out this means that in effect the operation of the GEG will, to some degree, reduce the amount of sun energy that gets into the Earth's atmosphere or down to the ground. This is a form of the Solar Radiation Management (SRM) type of geo-engineering.

Furthermore, a good deal of the electricity the GEG generates will never go down to the ground, as it will be used to run the HST system. This only makes the operation of the GEG and HST together even more effective as an SRM system.

If we add to each group of solar panels an additional radiator pointed at the ground below, and a valve to allow switching the cooling fluid coming off the solar panels to either the space-facing radiator or the ground-facing radiator, we would have a means to choose whether or not that amount of incoming solar energy would be delivered to the ground or would be sent out into space.

That gives us a way to control the overall amount of solar energy the Earth receives—in effect, putting a thermostat on the Earth. And in addition, it would allow us to make that control local to each region of the earth—creating not one, but a huge number of local thermostats for the planet.

And if the ground-facing radiators can be pointed either straight down or off to the side, we could control where on the Earth that solar energy is sent, at least to some degree. These capabilities might let us moderate the regional temperature differences that drive the weather patterns near the surface.

I pointed out above that we will only need to intercept a fraction of one percent of the incoming sunlight to generate all the electric energy we will need to replace all of our present uses of fossil fuels and any other sources of energy we will need for the foreseeable future. The number of panels that were needed to do that might be enough to also provide all the management of the solar insolation ("insolation" is the technical term for the amount of sunlight falling on the Earth) we'd need to accomplish whatever we wanted in terms of both local and global weather and climate control.

If we decide that we need to control more of the sunlight reaching the planet, we could do so by simply building more sunlight absorbers—mock solar panels if you will—and redirecting all of the energy they absorb just as we would the waste heat coming off of the solar panels. In that way we could reduce the insolation by perhaps as much as a couple of percent—which would be plenty to compensate for even the worst carbon dioxide pollution we are likely to have created by the time the GEG is built.

A Wholly Reversible Means of Solar Radiation Management

This is a form of Solar Radiation Management (SRM) that is completely controllable. We don't yet know how to use it, but as we experiment and learn we'll be able to immediately turn it up or down—or completely off—region by region, if we discover that it isn't helping us as we had hoped. This alone distinguishes this SRM approach from all the others I know about.

Deciding who gets to decide which areas are to get more or less solar energy input and when will raise many serious political issues that we'll have to deal with. But simply creating this new possibility would give humanity a whole new level of control over our global and regional environments. And that would, I believe, prove in the long run to be a real blessing.

I call this component of my proposal the **Global Climate and Regional Weather Control System** (or **GCRWCS** for short).

One additional comment: The ground-facing radiators are "looking at" a relatively warm surface (as opposed to the space-facing ones that see only the bitter cold of outer space). This means that they will absorb some earthshine even as they are radiating waste heat. This may necessitate having slightly larger radiators on the solar panel module, to keep the solar panels from overheating. They must be large enough to be sure that even with the earthshine the module picks up, the space-facing radiators can throw off enough waste heat in 24 hours to keep the solar panels at their optimum temperature.

For many locations the ground-facing radiators will be cooler than the ground below them. This means they will not be able to send out as much radiation as they absorb. Still, by sending down the amount that is usual for their temperature instead of having the module use its space-facing radiators to send that energy off into deep space, they will subtly warm the planet. You could think of this way: These modules will serve as partial reflectors of the earthshine, sending back some of the heat that would otherwise escape to space.

Part Six
Closing Comments

Additional Infrastructures for The Platform

Once The Platform is built, there are many things one might want to put on it. I'll just list a few here. I'm sure you can imagine quite a few others with just a little thought.

One could mount wind turbines on the lower reaches of The Platform, at about the altitude of the jet stream winds. This would let us do several desirable things: First, the wind turbines could generate more "green" energy for us. Second, they would be so high above the ground that no one would hear them, they wouldn't kill any flying birds, and no one on the ground could see them. Third, if we built enough of them they could extract enough energy from the jet stream to modify it, and therefore modify regional weather in a different way than I mentioned above. Like the Global Climate and Regional Weather Control System (GCRWCS) I just described, this would be a capability we have never had before. So we'd have to do many experiments to know what it could do before we could figure out what we wanted it to do. And, again, it would raise some serious political issues around who got to decide how that capability would be deployed.

The Platform would be an ideal place to mount astronomical telescopes. They would be held stably by its dynamic adjustment capability and they would have no atmosphere to distort the starlight, and no weather to interfere with their ability to see the stars. (Telescopes at ground level are often very heavy, to help keep them from being affected by local vibrations, etc. On top of The Platform, with its dynamic steadying, the telescopes could be reduced to their minimum optically necessary weight.)

The sides of The Platform trestles also would be an ideal place to mount antennas for radio and television stations and other telecommunications systems. They would be closer than satellites yet far enough up to reach broad areas of the planet. (Different uses of this sort would best be placed at different altitudes, and since The Platform would rise through all the possible altitudes the optimum altitude for each use could easily be accommodated.)

Real time observation of cities would be facilitated by mounting cameras on the lower reaches of The Platform where its segments passed overhead or nearby. This could be of great help to many segments of society, ranging from giving real-time traffic information to drivers to helping police and fire departments monitor their cities for problems that would need their services, etc.

Observations of farms could be equally helpful in letting farmers know when plants are suffering and need some attention. They now often use information from orbiting satellites for this, but by mounting permanent cameras above and not so far away from their farms they could get better data more of the time. A similar benefit would accrue to the managers of national and state parks, forests, and grasslands.

I described how the HST system could be used to support a garbage collection function in near-earth orbit. That doesn't directly address the problem of space rocks from the asteroid belt crashing into the Earth. There are other programs now being worked on to attempt to solve that. But some variation on it might work well and cost less than anything now being proposed.

One proposed program called DE-STAR is being developed by Philip Lubin and others at the University of California, Santa Barbara. This device would attempt to deflect any such space rock before it hits the Earth by shining a really powerful laser on it to boil away some of the rock, and as that vapor leaves the rock, the rock will recoil in the opposite direction, thus shifting its orbit. Their idea is to put these powerful lasers in orbit.

I suggest that they could do better by mounting the DE-STAR lasers on The Platform and powering them from the GEG. That would create a permanent installation that would be extremely stable and not moving nearly as fast as an orbiting device—thus making the aiming of the lasers easier and supplying the needed energy far easier. (The DE-STAR laser must, in any case, not be ground-based as it must be outside the atmosphere to avoid having the beam energy absorbed by the atmosphere, which not only would diminish its effectiveness in deflecting the space rock, but also might cause havoc in some fashion by the localized heating in our atmosphere.) Built on top of The Platform the DE-STAR system could also be far heavier and thus could have far more powerful lasers than anything we could launch into orbit with rockets, and it would cost far less to build on The Platform.

And the list could go on and on....

Some Thoughts About Governance and More

I've mentioned a few times that governance will be a serious issue. Any system that will affect every society on earth in so many significant ways will need a very smart governance arrangement.

I think that the energy generation and transportation components may be the easiest to govern. Pretty much everyone would benefit from having them work well. So there should be little conflict over the measures that would insure that outcome. Although deciding how to share a limited supply of energy (before the full system is built) would certainly be problematic.

For this sort of thing I think we can look to how the Internet is now governed. The United States built it at first, but now every nation has representatives on the two institutions that govern the Internet. They are the **Internet Corporation for Assigned Names and Numbers** (ICANN) that is responsible for the coordination of the global Internet's systems of unique identifiers and, in particular, ensuring its stable and secure operation,[37] and the **World Wide Web Consortium** (W3C) which is an international community that develops open engineering and other technical standards to ensure the long-term growth of the Web.[38]

Another international organization that has seemed to work for the good of all, rather than for any particular nation's benefit, is the World Health Organization. And, as with the Internet, we all benefit when diseases are minimized or eliminated. So it is quite natural that the WHO gets a lot of international cooperation, and that it generally shows no favor to any particular country or countries.

I believe that those who build (and pay for) the initial construction of The Platform and these two critical infrastructures (solar energy and transportation) built on top of The Platform should also create similar organizations to insure that all the technical and political issues surrounding the building and maintenance of these infrastructures is handled in a way that benefits all.

When it comes to the Global Climate and Regional Weather Control System (GCRWCS) and also any wind power system that is sufficiently large to affect weather or climate, a different model may be necessary. Here there may very well be clear winners and losers depending on how those systems are operated.

It would be prudent to spend some significant energy on figuring out how this sort of system should be managed well *before* we do the experiments that

---

[37] See https://en.wikipedia.org/wiki/ICANN.

[38] See http://www.w3.org/.

will let us know what impact they may have. I would not want to see the global corporations running these systems for their exclusive benefit. I'd far rather see some political system in place to protect the interests of the world's people above all else. (Otherwise, it might turn out—to give just one example—that the corporations would give preference to insuring good crops in, for example, the farmlands in the United States and Canada over allowing monsoons to develop over Bangladesh—with the result that the Bangladeshis who depend on their seasonal monsoons for survival would perish, simply because the corporations desired more profit from their farming enterprises.)

There are a host of other issues that need to be understood to assess the full importance of these proposed infrastructures. Again, you will find many of them discussed in my longer book (see footnote on page 50).

An Important Cautionary Statement

Someone will have to be in charge of building, operating, and maintaining these new global infrastructures. Who exactly will take the responsibility for how these activities are performed matters a great deal.

I envision these new infrastructures as being global not only in their physical extent, but that they will serve all of mankind. That is, they will be managed in a manner that best meets the needs of everyone.

This won't happen automatically. Indeed, if history is any guide, making that happen will be very difficult—yet vitally important. We must not let these infrastructures be run by and for the benefit of only certain wealthy and powerful interests. Certainly they must not be wholly controlled by large multinational corporations.

I caution that we, all the people who inhabit the world, must collectively insist that the governance organizations that are created for these global infrastructures be based on democratic principles. Everyone who has an interest in how they are built and run—and that is everyone on the planet— must be represented.

Those who now hold power usually seek to capture and control any new source of power (and, often, to deny that power to others). Make no mistake: These new infrastructures will be enormously powerful. They will impact the lives of everyone living in or near any urban center daily which now means more than half of all humanity, and by the middle of this century will be more than three-quarters of all humanity.

This is our challenge as We the People of all Humanity: Make sure that these new and extremely powerful infrastructures are created and run in a way that serves us all. Good luck! We'll need it.

Summary Comments

By the time you've gotten to this point you've become aware of just how large a project I am proposing, and also how very important it may be to our survival as a species.

Building it will be a serious engineering and political challenge. I expect that the initial research (to prove the feasibility and get some estimates on the best way to build The Platform and the probable cost for that construction) will be done in the United States, Europe, and Japan, with perhaps some help from Russia, China, and India.

The actual construction of the first segments of The Platform (and the solar energy and transportation infrastructures built on top of it) will likely take place in Russia or perhaps China teamed with Dubai, Saudi Arabia, or other countries in the Middle-East. I think Russia may be first because it is the country that spans the largest number of time zones. It could build segments of The Platform that span far enough around the planet to get solar power about twenty hours out of every day, at least in the summer. And they wouldn't need anyone else's permission to do that.

It and all the others I just mentioned are countries where a decision like this can be made by just a few top government officials. So they can make and implement that sort of huge project much more quickly than would be possible in a democratic country with all the political wrangling that would likely occur. (Saudi Arabia and Dubai, for example, also have this freedom of action plus the wherewithal to buy permission to build across some of their neighboring countries if they and China should choose to build segments that span from the Middle East to Eastern China in order to get a similar reach around the globe to what Russia could build, and one that is even nearer to the equator, and thus more useful all year long.)

But wherever that first segment is built, once its benefits are demonstrated, I think nearly every developed country will hurry to catch up. No one will want to be left out. As soon as they can, even the least developed nations will want in—and they will have every motivation to be sure they do get in, in no small part because if that is done right, it will lead to people changing their values in ways that will help stabilize the populations of those countries very much more quickly than would otherwise be the case. For why I say this, please see my companion book, *Multiple Population-Related Problems and a Surprisingly Graceful Solution* (see footnote on page 6).

Remember what has happened in the past when revolutionary new transportation (or energy generation) systems have been introduced. A lot of things happened, fitting into either one of two general categories:

1. Life was vastly improved; the economy boomed; people were able to do more, go farther from their homes, and generally had more options. These are the desirable outcomes.

2. While things started out very chaotic at first, soon the people who were in power before the revolution asserted their power and took control over the new capabilities, reducing the chaos, but also making for less personal freedom of choice and action. These are, in my view, some of the less-desirable outcomes.

Here are a few examples:

During the nineteenth century the United States built out its railroad system. This one act (carried out by a few large corporations with generous government support) changed the country's society and revolutionized its business practices. Similar revolutionary changes were made in many other countries around the world. Before the railroads, there were no time zones. And any place that was more than a few miles from one's home was essentially unreachable without dedicating one—or often many—days merely to getting from one place to another. Air transport, when it arrived, revolutionized things again, making it possible to cover huge distances in unthinkably small amounts of time, at least if you could afford to use an airplane to get there.

Starting in the 1950s the United States built the Interstate Highway system. It was meant to enhance the nation's ability to move people and cargo around the country quickly and safely. And it, too, it changed the entire society in many ways. Anyone whose residence or business was near an interstate highway exit found their life was vastly improved. Many others found their businesses lost patronage and their homes lost value. Indeed, many small cities that were not fortunate enough to be located on an interstate highway are no longer in existence—or they have been reduced to little more than a ghost town.

I think most of us would judge these past revolutions as having had overall a beneficial effect. Not many people would like to give up what they see as the benefits of those transformative changes in our societies. So if we do build these new infrastructures, I certainly hope we will be alert to the changes they will bring, and will do our best to see that the benefits far outweigh the drawbacks.

I don't want to minimize the magnitude of the engineering and manufacturing challenges that building The Platform presents. But at the same time I don't want anyone thinking that simply because it appears difficult and is something we've never done, that we cannot now do it.

* * *

Remember, please, what United States President John F. Kennedy said back on September 12, 1962, *"We choose to go to the moon. We choose to go to the moon in this decade and do the other things, not because they are easy, but because they are hard, because that goal will serve to organize and measure the best of our energies and skills, because that challenge is one that we are willing to accept, one we are unwilling to postpone, and one which we intend to win, and the others, too.* After he said that the country committed itself to the task and there was a palpable sense of excitement and optimism in the land. And we did accomplish all that he said we would, and we did it on time. We can create that same feeling for the entire world by choosing to build the infrastructures I am proposing.

For the past 50 years NASA has been pumping out technical innovations and giving them all away to the world—at no charge. What they have given us has transformed much of our society. I'm sure that 50 years after we commit to these new infrastructures, we'll be reaping a similar huge number of benefits aside from what those structures themselves will provide.

And remember also, these wise words of Arthur C. Clarke, the scientist and author who also was the inventor of the telecommunications satellite, *"New ideas pass through three periods:*
1) It can't be done.
2) It probably can be done, but it's not worth doing.
3) I knew it was a good idea all along."

If you have questions or comments and would like to share them with me, please do so by emailing me at this address: mailto:AvoidingOurExtinction@gmail.com

Acknowledgments

I must not end this without acknowledging those who helped me so very much: First is my wife, Pauline Merry, for her patience with all the time this has taken, and her *im*patience in urging me onward with all possible haste, since she could see how very important this project is. I thank my book agent, Bill Gladstone, and his company, Waterside Productions, for publishing this and for his enthusiasm and generously given wise counsel.

Finally, I'll just list the names of several others who helped me in ways large and small. You all know what you did, and I thank you profusely for it: Gregory Benford, Pat Burns, MichelJoy DelRe, Phil Feldhus, Albert Goodman, Keith Hjelmstad, Geoffrey Landis, John Lunsford, Herb Shapiro, Mike Sloan, Maki Takeichi, and Pam & Lee Willmore.

Each of these people helped me better understand the subject and/or reduce the number of errors and confusing or misleading statements. However, any mistakes or confusions that remain are solely my responsibility.

The cover background image is a screen capture from Google Earth. The background images and the magenta lines in the figures on pages 42 and 86 are also screen captures from Google Earth—in these cases when it was running a KML program I created to visualize the location of one segment of The Platform (shown as those magenta lines) in context. And the image of the Earth in the space elevator figure on page 51 is yet another Google Earth screen capture.

John M. Goodman, May, 2014
AvoidingOurExtinction@gmail.com

The Author

John M. Goodman is a writer, designer, consultant, and inventor.

His teaching career began at Harvey Mudd College, and later at the Joint Science Center of the Claremont Colleges where he was a Professor of Physics. At California Institute of the Arts he was the only faculty member teaching mathematics, science, and the history of science. Later he taught physics, computer science, and other subjects at California State University Fullerton, National University, and several community colleges and proprietary schools.

He served as a consultant to Scientific American magazine on two of their single-topic special issues (1967 and 1968). As a consultant to the Charles and Rae Eames design studio, and later for the Glenn Fleck design studio, he designed numerous interactive museum exhibits.

He was one of the founding faculty of the SELF alternative public high school in Irvine, CA. In 1978 he founded The Experience Center, a small interactive science museum in Irvine, CA and served as its Executive Director, a Trustee, and its exhibit designer. That institution closed in 1983, but with the use of some of its exhibits and several of its Trustees another organization in the county opened the Discovery Science Center in Santa Ana, CA in 1998.

He taught classes on various aspects of personal computer technology and became a course director for a seminar company teaching classes all over North America on the maintenance of personal computers. Out of this experience he began writing books for a general market explaining how personal computers work. He has seven published titles including best sellers such as "DOS 6.0 Power Tools" and "Peter Norton's Inside the PC" (Seventh and Eight Editions). He also has written numerous articles for popular and technical journals.

His honors and awards include being a winner in the Westinghouse Science Talent Search (now the Intel Science Search), one of the first class of National Merit Scholars at Swarthmore College, a Danforth Fellow and a National Science Foundation Fellow at Cornell University. His book "Hard Disk Secrets" was honored by the Computer Press Association and he earned a Francis W. Sears Award for excellence in demonstration apparatus design from the American Association of Physics Teachers for one of his novel designs for use by physics teachers. His work as an exhibit designer was honored by the American Association of Museums. He also has patents in diverse areas of technology.

He is a life member of Phi Beta Kappa and Sigma Xi, and has been a member of the American Association for the Advancement of Science, American Association of Museums, American Association of Physics Teachers, American Physical Society, Computer Press Association, Institute of Electrical and Electronic Engineers, Mensa, Museum Educators of Southern California, and the Orange County Arts Alliance.

His passions include theater, visual arts, modern dance, plus jazz and chamber music. He lives with his wife of 33 years, Pauline Merry, in Garden Grove, CA.

www.ingramcontent.com/pod-product-compliance
Lightning Source LLC
Chambersburg PA
CBHW071238020426
42333CB00015B/1525